P9-CLS-808

WEIRD MATH

ALSO BY DAVID DARLING

*The Universal Book of Mathematics:
From Abracadabra to Zeno's Paradoxes*

Teleportation: The Impossible Leap

*Gravity's Arc: The Story of Gravity, from
Aristotle to Einstein and Beyond*

Life Everywhere: The Maverick Science of Astrobiology

*Equations of Eternity: Speculations on
Consciousness, Meaning, and the Mathematical
Rules That Orchestrate the Cosmos*

WEIRD MATH

A TEENAGE GENIUS & HIS TEACHER REVEAL THE STRANGE CONNECTIONS BETWEEN MATH & EVERYDAY LIFE

DAVID DARLING &
AGNIJO BANERJEE

BASIC
BOOKS
New York

Basic Books
Hachette Book Group
1290 Avenue of the Americas, New York, NY 10104
www.basicbooks.com

Printed in the United States of America
First Edition: April 2018

Published by Basic Books, an imprint of Perseus Books, LLC, a subsidiary of Hachette Book Group, Inc. The Basic Books name and logo is a trademark of Hachette Book Group. The publisher is not responsible for websites (or their content) that are not owned by the publisher.

Print book interior design by Cynthia Young at Sagecraft Book Design.

Library of Congress Cataloging-in-Publication Data
Names: Darling, David J., author. | Banerjee, Agnijo, author.
Title: Weird math : a teenage genius and his teacher reveal the strange connections between math and everyday life / David Darling and Agnijo Banerjee.
Description: New York : Basic Books, [2018] | Includes bibliographical references and index.
Identifiers: LCCN 2018001764 | ISBN 9781541644786 (hardcover) | ISBN 9781541644793 (ebook)
Subjects: LCSH: Mathematics—Popular works.
Classification: LCC QA93 .D285 2018 | DDC 510—dc23
LC record available at https://lccn.loc.gov/2018001764

ISBNs: 978-1-5416-4478-6 (hardcover), 978-1-5416-4479-3 (ebook)

LSC-C

10 9 8 7 6 5 4 3 2 1

Go down deep enough in anything and you will find mathematics.

—DEAN SCHLICTER

The most incomprehensible thing about mathematics is that it is comprehensible.

—KIRAN MA

CONTENTS

PREFACE

Math *is* weird. Numbers go on forever—and there are different kinds of forever. Prime numbers help cicadas survive. A (mathematical) ball can be cut up and then put back together, without any gaps, to make a ball twice the size, or a million times the size, of the original. There are shapes that have fractional dimensions and curves that fill a plane, leaving no holes. While bored by a dull presentation, physicist Stanislaw Ulam wrote out numbers, starting from 0, in a spiral form, marked in all the prime numbers, and found that many primes lie on long diagonals—a fact that is still not fully unexplained.

We forget sometimes how weird math is because we're so used to dealing with what seem like ordinary numbers and calculations, the stuff we learn about in school or use every day. Yet the fact that our brains are so adept at thinking mathematically, and, if we choose, to doing really complex and abstract math, is surprising. After all, our ancestors, tens or hundreds of thousands of years ago, didn't need to solve differential equations or dabble in abstract algebra in order to stay alive long enough to pass on their genes to the next generation. While they searched for their next meal or a place to shelter, there was nothing to be gained from musing about geometry in higher dimensions or theories of prime numbers. Yet we're born with brains that have the potential to do these things and to uncover, with each passing year, more and more extraordinary truths about the mathematical universe. Evolution has provided us with this skill, but how and why? Why are we, as a species, so good at doing something that has every appearance of being just an intellectual game?

Somehow math is woven into the very fabric of reality. Dig deep enough, and we find that what seemed to be tangible bits of matter or energy—electrons or photons, for instance—dissolve into immateriality, becoming mere waves of probability, and all we're left with is a ghostly calling card in the form of some intricate but beautiful set of equations. In some sense, mathematics underpins the physical world around us, forming an invisible infrastructure. Yet it also goes beyond this, into abstract realms of possibility that may forever remain purely exercises of the mind.

We've chosen in this book to highlight some of the more extraordinary and fascinating areas of math, including those where exciting new developments are in the offing. In some cases, they have links with science and technology—particle physics, cosmology, quantum computers, and the like. In others, they represent, for now at least, math for math's sake and are adventures into an unfamiliar land that exists only in the mind's eye. We've chosen not to shy away from certain subjects just because they're hard. One of the challenges in describing many aspects of math for a general audience is that they're far removed from everyday experience. But in the end, some way can always be found to link what today's explorers and pioneers at the frontiers of mathematics are doing with the world of the familiar, even if the language we have to use isn't as precise as academics would ideally choose. It's perhaps true to say that if something, however obscure, can't be explained reasonably well to a person of normal intelligence, then the explainer needs to improve their understanding!

This book came about in an unusual way. One of us (David) has been a science writer for more than thirty-five years and has written many books on astronomy, cosmology, physics, and philosophy, even an encyclopedia of recreational math. The other (Agnijo) is a brilliant young mathematician and child genius, with an IQ of at least 162, according to Mensa, who, at the time of writing, has just finished training in Hungary in preparation for the 2017 International Mathematics Olympiad. Agnijo started coming to David for tuition in math and science at the age of twelve. Three years later, we decided to write a book together.

We sat down and brainstormed the topics we wanted to cover. David, for instance, came up with higher dimensions, the philosophy of math, and the math of music, while Agnijo was keen to write about large numbers (his personal passion), computation, and the mysteries of primes. Right from the start, we chose to lean toward anything unusual or downright weird and to connect this weird math, where possible, with real-world issues and everyday experience. We also made a commitment not to shy away from subjects just because they were tough, adopting as a mantra that if we can't explain something in plain language, then we don't properly understand it. David generally took on the historical, philosophical, and anecdotal aspects of each chapter, while Agnijo grappled with the more technical aspects. Agnijo fact-checked David's work, and David combined all the writing into finished chapters. It all worked surprisingly well! We hope you enjoy the result.

A NOTE TO THE READER

In glancing through the pages of this book, you may notice that it contains some symbols, including *x*'s, ω's (omegas), and even the odd ℵ (aleph). You'll find an occasional equation or an unfamiliar-looking combination of characters, such as 3↑↑3↑↑3 (especially in the chapters on large numbers and infinity). If you're a nonmathematician, don't be put off. They're just shorthand for ideas that, hopefully, we explain well enough in advance and thereby help us delve a little faster and deeper into the subject than would otherwise be possible. One of us (David) has taught math privately to students for many years and has yet to come across one who can't be good at it once they believe in themselves. The fact is that we're all natural mathematicians, whether we realize it or not. So, with that in mind, let's take the plunge. . . .

1

THE MATH BEHIND THE WORLD

> Even stranger things have happened; and perhaps the strangest of all is the marvel that mathematics should be possible to a race akin to the apes.
>
> —ERIC T. BELL

> Physics is mathematical not because we know so much about the physical world, but because we know so little; it is only its mathematical properties that we can discover.
>
> —BERTRAND RUSSELL

In terms of intellectual ability, *Homo sapiens* hasn't changed much, if at all, over the past one hundred thousand years. Put children from the time when woolly rhinos and mastodons still roamed the earth into a present-day school, and they would develop just as well as typical twenty-first-century youngsters. Their brains would assimilate arithmetic, geometry, and algebra.

And if they were so inclined, there would be nothing to stop them from delving deeper into the subject and someday perhaps becoming professors of math at Cambridge or Harvard.

Our neural apparatus evolved the potential to do advanced calculations, and understand such things as set theory and differential geometry, long before it was ever applied in this way. In fact, it seems a bit of a mystery *why* we have this innate talent for higher mathematics when it has no obvious survival value. At the same time, the reason our species emerged and endured is that it had an edge over its rivals in terms of intelligence and an ability to think logically, plan ahead, and ask "what if?" Lacking other survival skills, such as speed and strength, our ancestors were forced to rely on their cunning and foresight. A capacity for logical thought became our one great superpower, and from that, in time, flowed our ability to communicate in a complex way, to symbolize, and to make rational sense of the world around us.

Like all animals, we effectively do a lot of difficult math on the fly. The simple act of catching a ball (or avoiding predators or hunting a prey) involves solving multiple equations simultaneously at high speed. Try programming a robot to do the same thing, and the complexity of the calculations involved becomes clear. But the great strength of humans was their ability to move from the concrete to the abstract—to analyze situations, to ask if-then questions, to plan ahead.

The dawn of agriculture brought the need to accurately track the seasons, and the coming of trade and settled communities meant that transactions had to be carried out and accounts kept. For both of these practical purposes, calendars and business transactions, some kind of reckoning had to be developed, and so elementary math made its first appearance. One of the regions where it sprang up was the Middle East. Archaeologists have unearthed Sumerian clay trading tokens dating back to about 8000 BC that show that these people dealt with representations of number. But it seems that, at this early period, they didn't treat the concept as being separate from the thing being counted. For example, they had different-shaped tokens for different items, such as sheep or jars of oil. When a lot of tokens had to be exchanged between parties, the tokens were sealed inside containers called bullae, which had to be broken open to check the contents. Over time, markings began to appear on the bullae to indicate how many tokens were within. The symbolic representations then evolved into a written number system, while tokens became generalized for counting any kind of object and eventually morphed into an early form of coinage. Along the way, the concept of number became abstracted from the type of object being counted, so that, for example, five was five whether it referred to five goats or five loaves of bread.

The connection between math and everyday reality seems strong at this stage. Counting and record keeping are practical tools of the farmer and the merchant, and if these methods do the job, who cares about the philosophy behind it all? Simple arithmetic looks well rooted in the world "out there": one sheep plus one sheep is two sheep; two sheep plus two sheep is four sheep. Nothing could be more straightforward. But look more closely, and we see that already something a bit strange has happened. In saying "one sheep plus one sheep," there's the assumption that the sheep are identical or, at least for the purposes of counting, that any differences don't matter. But no two

The Egyptians had a good understanding of practical mathematics and put this to effect in the construction of the Pyramid of Khafre at Giza, shown here together with the Sphinx.

sheep are alike. What we've done is to abstract a perceived quality to do with the sheep—their "oneness," or apartness—and then operate on this quality with another abstraction, which we call addition. That's a big step. In practice, adding one sheep and one sheep might mean putting them together in the same field. But also in practice, the sheep are different and, digging a little deeper, what we call "sheep"—like anything else in the world—aren't really separate from the rest of the universe. On top of this, there's the slightly disturbing fact that what we take to be objects (such as sheep) "out there" are constructions in our brains built up from signals that enter our senses. Even if we grant that a sheep has some external reality, physics tells us that it's a hugely complicated, temporary assemblage of subatomic particles that's in constant flux. Yet somehow, in counting sheep we're able to ignore this monumental complexity or, rather, in everyday life not even be aware of it.

Of all subjects, mathematics is the most precise and immutable. Science and other fields of human endeavor are, at best, approximations to some ideal and are always changing and evolving over time. As German mathematician Hermann Hankel pointed out: "In most sciences, one generation tears down what another has built and what one has established another undoes. In mathematics alone each generation adds a new story to the old structure." From the outset, this difference between math and every other discipline is inevitable because math starts with the mind extracting what it recognizes as being most fundamental and constant among the messages it receives via the senses. This leads to the concepts of natural numbers, as a way of measuring quantity, and of addition and subtraction as basic ways of combining quantities. Oneness, twoness, threeness, and so on are seen as common features of collections of things, whatever those things happen to be and however different individuals of the same type of thing happen to be. So the fact that math has this eternal, adamantine quality to it is ensured from the start—and is its greatest strength.

Mathematics exists. Of that there's no doubt. Pythagoras's theorem, for instance, is somehow part of our reality. But where *does* it exist when it's not being used or being instantiated in some material form, and where *did* it exist many thousands of years ago, before anyone had thought about it?

Platonists believe that mathematical objects, such as numbers, geometric shapes, and the relationships between them, exist independently of us, our thoughts and language, and the physical universe. Quite what sort of ethereal realm they inhabit isn't specified, but it's a common assumption that they're somehow "out there." Most mathematicians, it's probably fair to say, subscribe to this school of thought and therefore also to the belief that math is discovered rather than invented. Most, too, probably don't care very much for philosophizing and are happy just to get on with doing math, in the same way that the majority of physicists, working in the lab or solving theoretical problems, don't worry a lot about metaphysics. Still, the ultimate nature of things—in this case of mathematical things—is interesting, even if we never arrive at a final answer. Prussian mathematician and logician Leopold Kronecker thought that only whole numbers were given, or in his words: "God made the integers, all the rest is the work of man." English astrophysicist Arthur Eddington went further: "The mathematics is not there till we put it there." The debate about whether mathematics is invented or discovered, or is perhaps some combination of both, arising from a synergy of mind and matter, will no doubt rumble on and, in the end, may have no simple answer.

One fact is clear: if a piece of math has been proven to be true, it will remain true for all time. There's no matter of opinion about it or subjective influence. "I like mathematics," remarked Bertrand Russell, "because it is not human and has nothing particular to do with this planet or with the whole accidental universe." David Hilbert voiced something similar: "Mathematics knows no races or geographic boundaries; for mathematics, the cultural world is one country." This impersonal, universal quality of math is its greatest strength yet doesn't, to the trained eye, detract from its aesthetic appeal. "Beauty is the first test: there is no permanent place in the world for ugly mathematics," remarked English mathematician G. H. Hardy. The same sentiment, but from the field of theoretical physics, was expressed by Paul Dirac: "It seems to be one of the fundamental features of nature that fundamental physical laws are described in terms of a mathematical theory of great beauty and power."

The flip side to the universality of math, however, is that it can seem cold and sterile, devoid of passion and feeling. As a result, we may find that although intelligent beings on other worlds share the same mathematics as us, it isn't the best way to communicate with them about a lot of the things that matter to us. "Many people suggest using mathematics to talk to the aliens," commented SETI (search for extraterrestrial intelligence) researcher Seth Shostak. In fact, the Dutch mathematician Hans Freudenthal developed an entire language (Lincos) based on this idea. "But," said Shostak, "my personal opinion is that mathematics may be a hard way to describe ideas like love or democracy."

The ultimate goal of scientists, certainly of physicists, is to reduce what they observe in the world to a mathematical description. Cosmologists, particle physicists, and the like are never happier than when they have measured and quantified things and then found a relationship between the quantities. The idea that the universe is mathematical at its core has ancient roots, stretching back at least as far as the Pythagoreans. Galileo saw the world as a "grand book" written in the language of mathematics, and, much more recently, in 1960, Hungarian-American physicist and mathematician Eugene Wigner wrote a paper called "The Unreasonable Effectiveness of Mathematics in the Natural Sciences."

We don't see numbers directly in the real world, so it isn't immediately obvious that math is all around us. But we do see shapes—the near-spherical shape of planets and stars, the curved path of objects when thrown or in orbit, the symmetry of snowflakes, and so on—and these can be described by relationships between numbers. Other patterns, translatable into math, emerge from the way electricity or magnetism behaves, galaxies rotate, and electrons operate within the confines of atoms. These patterns, and the equations describing them, underpin individual events and seem to represent deep, timeless truths underlying the changing complexity in which we find ourselves. German physicist Heinrich Hertz, who first conclusively proved the existence of electromagnetic waves, remarked: "One cannot escape the feeling that these mathematical formulas have an independent existence and an intelligence of their own, that they are wiser than we are, wiser even than their discoverers, that we get more out of them than was originally put into them."

It's unquestionably true that the bedrock of modern science is mathematical in nature. But that doesn't necessarily mean that reality itself is fundamentally mathematical. Ever since the time of Galileo, science has separated the subjective from the objective, or measurable, and focused on the latter. It's done its best to evict anything to do with the observer and to pay attention only to what it assumes lies beyond the interfering influence of the brain and senses. The way modern science has developed almost guarantees that it will be mathematical in nature. But this leaves

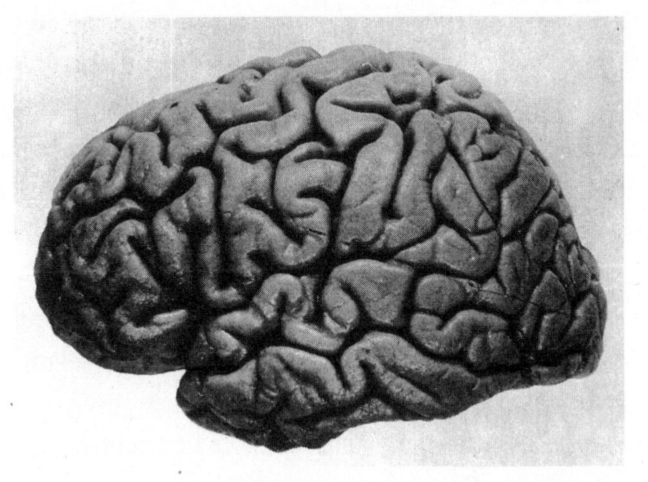

Why has the human brain evolved to be so extraordinarily good at a subject—mathematics—that it doesn't need for survival?

much that science has trouble dealing with—most obviously consciousness. It may be that someday we'll have a good, comprehensive model of how the brain works, in terms of memory, visual processing, and so forth. But why we also have an inner experience, a feeling of "what it is like to be," remains—and may always remain—outside the field of conventional science and, by extension, of mathematics.

On the one hand, Platonists believe math to be a land that already exists, awaiting our exploration of it. On the other, there are those who insist that we invent mathematics as we go along, to suit our purposes. Both positions have weaknesses. Platonists struggle to explain where things like pi might be outside the physical universe or the intelligent mind. Non-Platonists have a hard time denying the fact that, for example, planets would continue to orbit the sun in ellipses whether we do the math or not. A third school of mathematical philosophy occupies a middle ground between the two, by pointing out that, in describing the real world, math is not as successful as it's sometimes made out to be. Yes, equations are useful in telling us how to navigate a spacecraft to the moon or Mars, design a new aircraft, or predict the weather several days in advance. But these equations are mere approximations of the reality of what they're intended to describe, and, moreover, they apply to just a small portion of all the things going on around us. In touting the success of mathematics, the realist would say, we downplay the vast majority of phenomena that are too complex or poorly understood to capture in mathematical form or that, by their very nature, are irreducible to this kind of analysis.

Is it possible that the universe isn't, in reality, mathematical? After all, space and the objects it contains don't directly present anything mathematical to us. We humans rationalize and make approximations in order to model aspects of the universe. In doing so, we find mathematics extremely useful in enabling us to understand it. That doesn't necessarily imply that math is anything other than a convenience of our making. But if mathematics isn't present in the universe to start with, how is it that we're able to invent it in order to put it to such use?

Mathematics is broadly divided into two areas: pure and applied. Pure math is math for math's sake. Applied mathematicians put their subject to work on real-world problems. But often, developments in pure math, with seemingly no bearing on anything tangible, have turned out later to be surprisingly useful to scientists and engineers. In 1843 Irish mathematician

William Hamilton hatched the idea of quaternions—four-dimensional generalizations of ordinary numbers of no practical interest at the time but that, more than a century later, have turned out to be an effective tool in robotics and computer graphics and games. A question first tackled by Johannes Kepler in 1611, about the most efficient way to pack spheres in three-dimensional space, has been applied to the efficient transmission of information over noisy channels. The purest mathematical discipline, number theory, much of which was thought to have little practical value, has led to important breakthroughs in the development of secure ciphers. And the new geometry pioneered by Bernhard Riemann that dealt with curved surfaces proved ideal for the formulation of Einstein's general theory of relativity—a new theory of gravity—more than fifty years later.

In July 1915, one of the greatest scientists of all time met one of the greatest mathematicians of the age when Einstein paid a visit to David Hilbert at the University of Göttingen. The following December, both published, almost simultaneously, the equations that described the gravitational field of Einstein's general theory. But whereas the equations themselves were the goal for Einstein, they were what Hilbert hoped would be a stepping-stone toward an even grander scheme. Hilbert's passion, the driving force behind much of his work, was a search for fundamental principles, or axioms, that might underlie all of mathematics. Part of this quest, as he saw it, was to find a minimum set of axioms from which he could deduce not only the equations of Einstein's general theory but any other theory in physics as well. Kurt Gödel, with his incompleteness theorems, undermined faith in the notion that math might have the answers to all questions. But we remain uncertain as to what extent the world in which we live is truly mathematical or just mathematical in appearance.

Whole swaths of mathematics may never be put to use, other than to help open up yet more avenues of pure research. On the other hand, for all we know, it may be that much of pure math is enacted, in unexpected ways, in the physical universe—or if not this universe, then in others that might exist throughout what cosmologists suspect is a multiverse of incomprehensible scale. Perhaps everything that is mathematically true and valid is represented somewhere, sometime, somehow in the reality in which we are embedded. For now there is the journey to keep us occupied: the weird and wonderful adventure of the human mind as it explores further the frontiers of number, space, and reason.

In the chapters that follow, we'll take a deep dive into subjects that are both bizarre and astonishing yet, at the same time, have very real connections with the world we know. True, some of the math may seem esoteric, fanciful, and even pointless, like some strange and convoluted game of the imagination. But at its core, mathematics is a practical affair, rooted in commerce, agriculture, and architecture. Although it has developed in ways that our ancestors could never have dreamed of, still those links with our everyday lives remain at its heart.

2

HOW TO SEE IN 4-D

> One of the strangest features of string theory is that it requires more than the three spatial dimensions that we see directly in the world around us. That sounds like science fiction, but it is an indisputable outcome of the mathematics of string theory.
>
> —BRIAN GREENE

We live in a world of three dimensions—up and down, side to side, and backward and forward, or any other three directions that are at right angles to each other. It's easy to imagine something in one dimension, such as a straight line, or two dimensions, such as a square drawn on a sheet of paper. But how can we possibly learn to see in an extra dimension than those we're familiar with? Where is this additional direction that's perpendicular to the three we know?

These questions may seem purely academic. If our world is three-dimensional, why worry about 4-D or 5-D and so forth? The fact is that science may need higher dimensions to explain what is going on at a subatomic level. These extra dimensions may hold the key to understanding the grand scheme of matter and energy. Meanwhile, on a more practical level, if we could learn to see in 4-D, we would have a powerful new tool to deploy in medicine and education.

Sometimes the fourth dimension is taken to be something other than an extra direction in space. After all, the word "dimension," from the Latin *dimensionem*, means simply a "measurement." In physics the basic dimensions that form the building blocks of other quantities are considered to be length, mass, time, and electric charge. Very often, in a different context, physicists talk about three dimensions of space and one of time, especially since Albert Einstein showed that, in the world in which we live, space and time are always bound up together in a single entity called spacetime. Even before the theory of relativity came along, however, there had been speculation about the possibility of being able to move backward and forward along the dimension of time just as we can move any way we like in space. In his novel *The Time Machine*, published in 1895, H. G. Wells explains that an instantaneous cube, for instance, can't exist. A cube that we see moment by moment is just a cross section of a four-dimensional thing having length, breadth, width, *and duration*. "There is no difference," says the Time Traveller, "between Time and any of the three dimensions of Space except that our consciousness moves along it."

The Victorians were also fascinated by the idea of a fourth dimension of space, both from a mathematical point of view and in the possibilities it seemed to offer of explaining another obsession of the age—spiritualism. The late 1800s was a period when many people, including luminaries such as Arthur

Conan Doyle, Elizabeth Barrett Browning, and William Crookes, were attracted to the claims of mediums and the prospect of communicating with the dead. Might the afterlife, people wondered, exist in a fourth dimension that was parallel to, or overlapped with, our own so that spirits of the deceased could pass easily into our material realm and back again?

Our failure to be able to visualize in higher dimensions makes it tempting to think that the fourth dimension is somehow mysterious or alien to anything we know. Mathematicians, though, have no trouble in working with four-dimensional objects or spaces because they don't have to imagine what they're actually like in order to describe their properties. These properties can be figured out using algebra and calculus without having to resort to any multidimensional mental gymnastics. Start with a circle, for instance. A circle is a curve made of all the points in a plane that are at the same distance (the radius) from a given point (the center). Like a straight line, it has only length—no width or height—and so is a one-dimensional thing. Imagine yourself positioned and constrained within a line. The only freedom of movement you would have is along the line, one way or the other. It's the same with a circle. Although a circle exists in a space of at least two dimensions, if you were positioned and confined within the circle then you would have no more or less freedom of movement than if you were positioned within a line. You could only go back and forth along the circle, effectively tied down to a single dimension of movement.

Nonmathematicians sometimes think of a circle as including its interior as well. But a "filled-in circle" to a mathematician isn't a circle at all but a very different object called a disk.

A circle is a one-dimensional object that can be "embedded" in a two-dimensional object, a plane (a finely drawn circle on a sheet of paper is an approximation of this). The length, or circumference, of a circle is given by $2\pi r$, where r is the radius and the area enclosed by the circle is πr^2. Moving up a dimension, we come to the sphere, which consists of all the points lying at the same distance in three-dimensional space from a given point. Again, the layperson may confuse an actual sphere, which is just a two-dimensional surface, with the object that also includes all the points inside this surface. But once more, mathematicians make a sharp distinction and call the latter a "ball." A sphere is a two-dimensional object that can be embedded in three-dimensional space. It has a surface area of $4\pi r^2$ and encloses a volume of $\frac{4}{3}\pi r^3$. Because an ordinary sphere is two-dimensional, mathematicians call it a 2-sphere, whereas a circle, using the same naming system, is a 1-sphere. Spheres in higher dimensions are said to be "hyperspheres" and can be labeled in the same way. The simplest hypersphere, the 3-sphere, is a three-dimensional object embedded in four-dimensional space. We can't capture this in our mind's eye, but we can understand it by analogy. Just as a circle is a curved line and an ordinary (2-) sphere is a curved surface, a 3-sphere is a curved volume. Using some straightforward calculus, mathematicians can show that this curved volume is given by $2\pi^2 r^3$. It's the 3-sphere equivalent of the surface area of an ordinary sphere and is also referred to as a cubic hyperarea or surface volume. The four-dimensional space enclosed by a 3-sphere has a four-dimensional volume, or quartic hypervolume, of $\frac{1}{2}\pi^2 r^4$. Proving these facts about the 3-sphere is not much more difficult than proving them about the circle or ordinary sphere and doesn't involve having to understand what a 3-sphere actually looks like.

In the same way, we may struggle to grasp the true appearance of a four-dimensional cube, or "tesseract," though, as we'll see, we can try to represent it in two or three dimensions. But it's straightforward to describe the progression from square to cube to tesseract: a square has four vertices (corners) and four edges; a cube has eight vertices, twelve edges, and six faces; a tesseract has sixteen vertices, thirty-two edges, twenty-four faces, and eight "cells"—the three-dimensional equivalents of faces—consisting of cubes. This last fact is the one that defies our attempts at visualization: a tesseract has eight cubic cells arranged in such a way as to enclose a four-dimensional space, just as a cube has six square faces arranged so as to enclose a three-dimensional space.

The best we can normally do in coming to terms with the fourth dimension is to draw analogies with the third. For instance, if we were to ask, "What would a four-dimensional hypersphere look like if it were to pass through our space?" we can get an impression by considering what happens if a sphere passes through a plane. Suppose there are two-dimensional beings who inhabit that plane. Looking along the surface of their world—which is all they can do—they see merely dots or lines of different length that they can only interpret as two-dimensional figures. As our 3-D sphere initially makes contact with their 2-D space, they see it as a dot, which then grows into a circle, reaching a maximum diameter equal to the diameter of the sphere before the circle shrinks again to a dot and then disappears as the sphere passes through. Likewise, if a 4-sphere were to intersect our space, we would see it as a dot that expanded, like a bubble, into a three-dimensional sphere of maximum size before shrinking and finally vanishing. The true nature—the extra dimensionality—of the 4-sphere would be hidden from us, although its mysterious appearance, growth, and disappearance would probably cause us to wonder what was going on!

Four-dimensional beings would have seemingly magical powers in our world. They could, for example, pick up a right-footed shoe, flip it over in the fourth dimension, and put it back as a left-footed one. If this seems hard to understand, think of a two-dimensional shoe, which would be like an infinitely thin sole shaped for one foot or the other. We could cut such a shape out of a piece of paper, lift it up, turn it over, and put it back down, so that we changed its footedness. A 2-D creature would find this utterly astonishing, but to us, with the benefit of the extra dimension, the trick would seem obvious.

In principle, a 4-D being could flip around a whole (3-D) person in the fourth dimension, although the absence of cases of people having suddenly had everything switched right to left or left to right suggests that this hasn't actually happened. In his short tale "The Plattner Story," H. G. Wells describes the remarkable case of Gottfried Plattner, a teacher who disappears for nine days following an explosion in a school chemistry lab. Upon his return, he is effectively a mirror image of his previous self, though his recollections of what happened during the period of absence are met with incredulity. Being flipped over for real in the fourth dimension would be bad for your health, apart from the shock of seeing yourself look different in the mirror (faces are surprising asymmetrical). Many of the crucial chemicals in our bodies, including glucose and most amino acids, have a certain handedness. Molecules of DNA, for example, which take the form of a double helix, always twist like a right-handed screw. If all these chemicals had their handedness reversed, we would quickly die of malnutrition because much of the essential nutrients in our food, from plants and animals, would now be in a form we couldn't assimilate.

Mathematical interest in a fourth spatial dimension began in the first half of the nineteenth century with the work of German mathematician Ferdinand Möbius. He's best remembered for his study of a shape that's now named after him—the Möbius band—and as a pioneer of the field known as topology. It was he who first realized that in a fourth dimension, a three-dimensional form could be rotated into its mirror image. In the second half of the nineteenth century, three mathematicians stood out as explorers of the new realm of multidimensional geometry: Swiss Ludwig Schläfli, Englishman Arthur Cayley, and German Bernhard Riemann.

Schläfli began his magnum opus, *Theorie der Vielfachen Kontinuität* (Theory of Continuous Manifolds), by saying, "The treatise . . . is an attempt to found and develop a new branch of analysis that would, as it were, be a geometry of n dimensions, containing the geometry of the plane and space as special cases for $n = 2, 3$." He went on to describe multidimensional analogues of polygons and polyhedrons, which he called "polyschemes." These are now commonly known as polytopes, a term coined by German mathematician Reinhold Hoppe and introduced to English researchers by Alicia Boole Stott, daughter of English mathematician and logician George Boole, who devised Boolean algebra, and Mary Everest Boole, a self-taught mathematician and writer on the subject (and George's wife).

Also to Schläfli's credit is the discovery of the higher-dimensional relatives of the Platonic solids. By Platonic solid we mean a convex shape (one with all the corners pointing outward) with regular polygon faces and the same number of faces meeting at each corner. There are five of them: the cube, tetrahedron, octahedron, (twelve-sided) dodecahedron, and (twenty-sided) icosahedron. The four-dimensional equivalents of the Platonic solids are the convex regular 4-polytopes (also called polychora), of which Schläfli found there were six, named after the number of cells they have. The simplest 4-polytope is the 5-cell, which has 5 tetrahedral cells, 10 triangular faces, 10 edges, and 5 vertices and is analogous to the tetrahedron. Then there is the 8-cell, or tesseract, and its "dual," the 16-cell, obtained by replacing cells with vertices, faces with edges, and vice versa. The 16-cell has 16 tetrahedral cells, 32 triangular faces, 24 edges, and 8 vertices and is the four-dimensional analogue of the octahedron. Two other 4-polytopes are the 120-cell, an analogue of the dodecahedron, and the 600-cell, an analogue of the icosahedron. Finally, there is a 24-cell, which has 24 octahedral cells and no three-dimensional counterpart. Interestingly, Schläfli found, the number of convex regular polytopes in all higher dimensions is the same—just three.

Through the work of Cayley, Riemann, and others, mathematicians learned how to do complex algebra in 4-D and branch out into multidimensional geometries that went beyond the rules prescribed by Euclid. But what they still couldn't do was actually see in four dimensions. Could anybody? This was a problem that intrigued British mathematician, teacher, and writer of scientific romances Charles Howard Hinton. In his twenties and early thirties, Hinton taught at two private schools in England: first at Cheltenham College in Gloucestershire and then at Uppingham School in Rutland, where a fellow teacher—in fact, Uppingham's first mathematical master—was Howard Candler, a friend of Edwin Abbott. It was during this period, in 1884, that Abbott published his now classic satirical novel *Flatland: A Romance of Many Dimensions*. Four years earlier, Hinton had penned an article of his own on alternative spaces called "What Is the Fourth Dimension?" in which he put forward the idea that particles moving around in three dimensions might be thought of as successive cross-sections of lines and curves existing in four dimensions. We, ourselves, might really be four-dimensional beings "and our successive states the passing of them through the three-dimensional space to which our consciousness is confined." The extent of the relationship between Abbott and Hinton isn't clear, but they certainly knew of each other's work (and acknowledged as much in their writings), and some social contact would have taken place, if only via their mutual friend and colleague. Candler would surely have discussed with Abbott the young teacher at Uppingham who wrote and spoke so openly about other dimensions.

Hinton was nothing if not unconventional. At the time he was teaching in England, he married Mary Ellen Boole, daughter of the above-mentioned Mary Everest Boole (herself the niece of George Everest, after whom the tallest mountain is named) and George Boole. Unfortunately, three years into his marriage, Hinton also went through a secret wedding ceremony with another woman, Maud Florence, whom he had met while at Cheltenham College and with whom he had twin children. Probably the attitudes of his father, James

Cover of the first edition of Edwin Abbott's *Flatland*. (LONDON: SEELEY, 1884)

Hinton, a surgeon and head of a sect devoted to polygamy and free love, played a part in Charles's behavior. In any event, Hinton was found guilty of bigamy at an Old Bailey trial and jailed for several days. With his (first) family, he then fled to Japan, where he taught for some years, before becoming an instructor of mathematics at Princeton University. There, in 1897, he designed a species of baseball gun, which, with the help of gunpowder charges, fired out balls at speeds of forty to seventy miles per hour. The *New York Times* in its March 12 edition of that year described it as "a heavy cannon, with a barrel about two and a half feet in length, with a rifle attachment in the rear." Its cleverest trick, throwing curveballs, was accomplished with the help of "two curved rods, which are inserted in the barrel of the cannon." For a few seasons, the Princeton Nine used it, on and off, before abandoning it as a safety hazard. Whether the injuries it caused were a factor in Hinton's dismissal from the college is unclear, but they didn't prevent him from reintroducing the machine at the University of Minnesota, where, briefly, in 1900, he held a teaching post before joining the US Naval Observatory in Washington, DC.

Hinton's fascination with the fourth dimension, stretching back to his early days as a teacher in England, began at a time when others were writing about the subject and often speculating about its possible links with spiritualism. In 1878 Friedrich Zöllner, professor of astronomy at the University of Leipzig, published a paper called "On Space of Four Dimensions" in the *Quarterly Journal of Science* (edited by chemist and prominent spiritualist William Crookes). Zöllner started on solid mathematical ground by referencing Bernhard Riemann's seminal paper "On the Hypotheses Which Underlie Geometry," published in 1868, two years after Riemann's death and fourteen years after its contents were first delivered as a lecture by Riemann while still a student at the University of Göttingen. Riemann developed the concept, first hinted at by his supervisor at Göttingen, the great Carl Gauss, that three-dimensional space could be curved (just as a two-dimensional surface, such as a sphere, can be) and extended this idea of the curvature of space into an arbitrary number of dimensions. The result, known as elliptic or Riemannian geometry, later formed a cornerstone of Albert Einstein's general theory of relativity. Zöllner also borrowed the notion, described in an 1874 paper by the young projective geometer Felix Klein, that knots could be undone and rings unlinked simply by lifting them into a fourth dimension and turning them over. In this way, Zöllner set the scene for his explanation of how spirits, existing, as he saw it, on a higher plane, could perform the various phenomena—especially the knot-untying tricks—that he had witnessed at séance experiments with the famous (and, as it turned out, utterly fraudulent) medium Henry Slade. Hinton, like Zöllner, was inclined to think that mere habit of perception limited us to a three-dimensional viewpoint and that a fourth dimension might be all around us and become visible to us if only we could train ourselves to see it.

Although something that's four-dimensional is hard to imagine, it's easy to do a 2-D sketch of one. This is especially true in the case of the four-dimensional equivalent of a cube for which Hinton coined the name "tesseract." Start by drawing two squares, slightly offset and connecting their corners by straight lines. This can be visualized as a perspective drawing of a cube, the squares being separated, in our mind's eye, in the third dimension. Next draw two cubes joined at their corners. With 4-D vision we would be able to see this as two cubes separated in the fourth dimension—in fact, a perspective of a tesseract. Unfortunately, flat representations of 4-D objects aren't much help to us in being able to see them for what they really are. Hinton realized that a more fruitful approach to training our minds to see in four dimensions might be through three-dimensional models that could be rotated to show different aspects of a 4-D shape: at least that way we would be dealing only with a perspective of the real thing rather than a perspective of a perspective. To this end, he developed an intricate visual aid in the form of a set of one-inch wooden cubes in different colors. A complete set of Hinton cubes consisted of eighty-one cubes painted in sixteen different colors; twenty-seven "slabs" used to represent, by analogy, how a 3-D object can be built up in two dimensions; and twelve multicolored "catalogue cubes." By elaborate manipulations, described in detail in his book *The Fourth Dimension*, first published in 1904, he was able to represent the various cross-sections of a tesseract and then, by memorizing the cubes and their many possible orientations, gain a window into this higher-dimensional world.

Did Hinton actually learn to create four-dimensional images in his brain? In addition to the familiar up and down, forward and back, and side to side, could he see "kata" and "ana"—his names for the two opposite directions along the fourth dimension? Without getting inside his head, we can't know. Certainly, he wasn't alone in building 3-D representations of 4-D shapes. He introduced his cubes to his sister-in-law Alicia Boole Stott, who became an intuitive geometer of the fourth dimension herself and adept at making card models of 3-D cross-sections of 4-D polytopes. The question remains whether, by such means, a person can develop true four-dimensional vision or just the ability to understand and appreciate the geometry of higher-dimensional objects.

In a way, being able to see an extra dimension is like being able to see a new color—one outside all our previous experience. French impressionist painter Claude Monet underwent surgery in 1923, at the age of eighty-two, to remove the lens from his left eye, which had become hopelessly clouded by cataracts. Subsequently, the colors he chose to use in his art changed from mostly reds, browns, and other earthy tones to blues and violets. He even repainted some of his earlier works so that, for example, what had been white water lilies took on a bluish hue—an

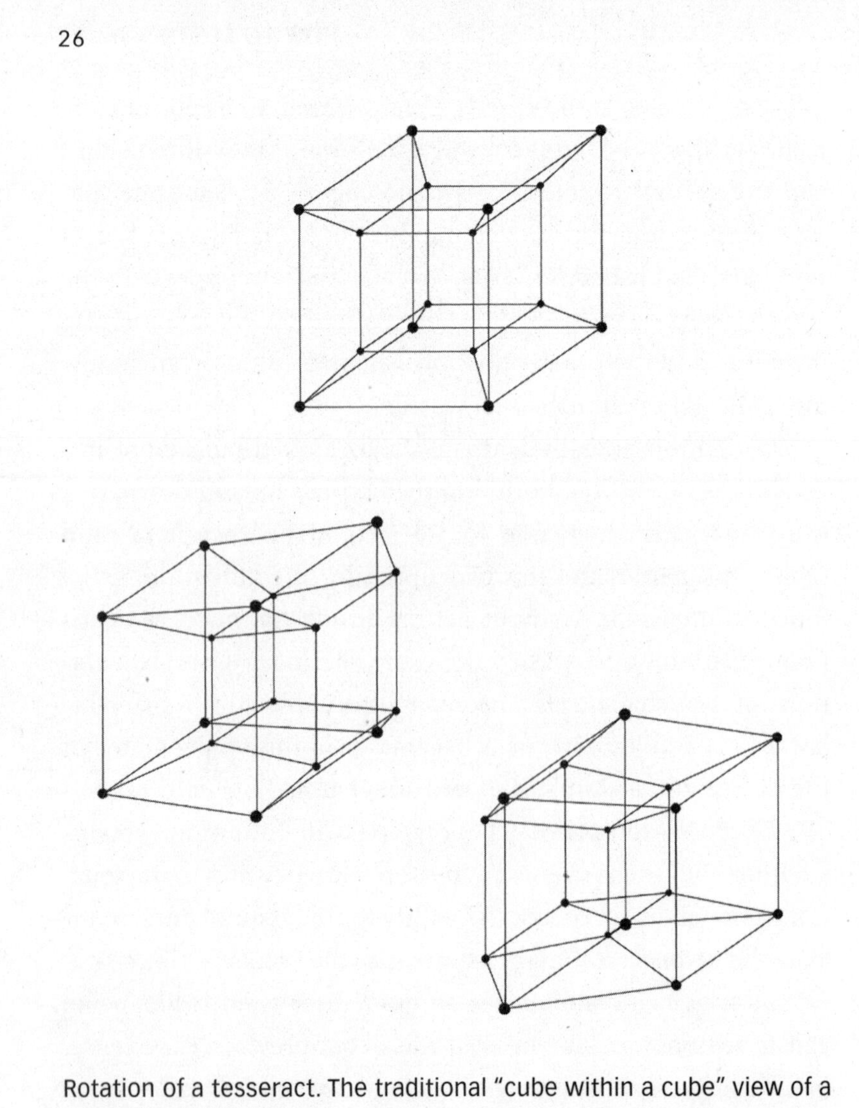

Rotation of a tesseract. The traditional "cube within a cube" view of a tesseract (*top*). The tesseract has rotated slightly. The central cube has started to move and is in the process of becoming the right cube (*center*). The tesseract has rotated farther, and the central cube is now much closer to where the right cube was originally (*bottom*). Finally, the tesseract rotates fully back to its starting position. What is important is that the tesseract has not in any way deformed. Instead, the changes are due to a shift in perspective. (AGNIJO BANERJEE)

indication, it's been claimed, that he could now see into the ultraviolet region of the spectrum. This idea is supported by the fact that the lens of the eye blocks out wavelengths shorter than about 390 nanometers (billionths of a meter), at the far end of the violet range, even though the retina has the potential to detect wavelengths down to about 290 nanometers, which is in the ultraviolet. There's also plenty of evidence, in more recent times, of young children and of older people who have missing lenses being able to see beyond the violet end of the spectrum. One of the best-documented cases is that of a retired air force officer and engineer from Colorado, Alek Komarnitsky, who had a cataract-affected natural lens replaced by an artificial one that can transmit some UV light. In 2011 Komarnitsky underwent tests using a monochromator at a Hewlett-Packard lab, where he reported being able to see wavelengths down to 350 nanometers as a deep purple hue and some variation in brightness even farther into the UV, down to 340 nanometers.

Most of us have three types of cone cells—the kind responsible for color vision—in our retinas. Most color-blind people, and many other types of mammals, including dogs and New World monkeys, have only two, so that the number of different shades of color they can see is restricted to about 10,000 compared to the 1 million or so the rest of us can discern. Researchers have, however, found rare instances of individuals with four different working types of cone cells. These "tetrachromats" can, according to estimates, distinguish almost 100 million times more shades of color than normal, although, since it's natural for everyone to assume we all see the same, they may only gradually come to realize that they have this superpower, without special testing.

The point is that humans have the ability, in special circumstances, to see things that are outside the normal experiences that most of us have. If some people can see in ultraviolet, or subtler shades than usual, then why not the fourth dimension? Evidently, our brains can adapt to processing sensory information that we're not normally used to receiving. Perhaps they can be trained to create internal images that are in 4-D.

Today, we have a huge advantage in our efforts to visualize the world of four dimensions, thanks to the availability of computers and other advanced technology. It's easy now to create animations of a wire-framed tesseract, for example, to show how its appearance, as seen on a flat screen, changes as it's rotated. Our brains still interpret what we see as the strange behavior of a number of interconnected cubes rather than anything in 4-D. Yet we get an impression of something very unusual going on that can't be explained in ordinary three-dimensional terms. Does the technology we have, or soon will have, hold the promise of letting us directly experience the fourth dimension?

One school of thought says that, despite the claims of people like Hinton, we can never really see in 4-D because the world around us is unremittingly three-dimensional, our brains are three-dimensional, and evolution has equipped us to interpret all the sensations we receive as set in a 3-D context. No amount of mental effort will help bring the particles that make up our bodies into a different plane of existence. Nor will any trick of engineering allow us to build a thing in 4-D, such as an actual tesseract. This hasn't stopped science fiction writers from imagining some strange combination of events that might cause a 3-D object or system to spontaneously develop an extra dimension. "And He Built a Crooked House," by Robert Heinlein, first published by *Astounding Science Fiction* in February 1941, tells the tale of an ingenious architect who designs a house with eight cubical rooms laid out like a net of a tesseract in 3-D. Unfortunately, an earthquake shakes the building, shortly after its completion, and causes it to fold into an actual hypercube, with bewildering results for those who first venture through its door. In "A Subway Named Moebius" (1950), Boston's underground train network becomes so convoluted that part of it flips into another dimension along with a train full of passengers, although all arrive safely at their intended stations in the end. Written by A. J. Deutsch, an astronomer at Harvard (one of the stops on the system), it plays on the themes of the Möbius band and Klein bottle, the latter being a one-sided shape that can exist only in four dimensions.

Artists too have tried to capture the essence of 4-D in their work. In his 1936 *Dimensionist Manifesto*, Hungarian poet and art theorist Charles Tamkó Sirató claimed that artistic evolution had led to "Literature leaving the line and entering the plane . . . Painting leaving the plane and entering space . . . [And] sculpture stepping out of closed, immobile forms." Next, Sirató said, there would be "the artistic conquest of four-dimensional space, which to date has been completely art-free." Salvador Dalí's *Crucifixion (Corpus Hypercubus)*, completed in 1954, unites a classical portrayal of Christ with an unfolded tesseract. In a 2012 lecture given at the Dalí Museum, geometer Thomas Banchoff, who advised Dalí on mathematical issues connected with his paintings, explained how the artist was trying to use "something from a three-dimensional world and take it beyond. . . . The exercise of the whole thing was to do two perspectives at once—two superimposed crosses." Dalí, like the nineteenth-century scientists who tried to rationalize spiritualism in terms of existence in some higher space, used the idea of the fourth dimension to connect the religious with the physical.

Twenty-first-century physicists have a new reason to be interested in higher dimensions: string theories. Here, subatomic particles, such as electrons and quarks, are treated as being not point-like but one-dimensional vibrating "strings." One of the strangest aspects of string theories is that, in order to be mathematically consistent, they require that the space and time in which we live have extra dimensions. A version called superstring theory calls for a total of ten dimensions, an extension of this known as M-theory involves eleven, while another scheme by the name of bosonic string theory demands twenty-six. All of these additional dimensions are said to be "compactified," meaning that they're significant only on a fantastically small scale. Maybe someday we'll learn how to amplify or uncurl these dimensions or observe them as they actually are. But for now and the foreseeable future, we're stuck with our familiar three macroscopic dimensions of space. So, the question remains: Is there any way we can visualize, in our minds, what a four-dimensional object is really like?

Our visual experience of the world comes about from light entering our eyes, striking our retinas, and creating two flat images. The light-sensitive cells in the retina generate electrical signals, which travel to the visual cortex in the brain where a 3-D reconstruction takes place based on essentially 2-D information. Having two eyes means that we can see objects from two slightly different angles, and the brain learns, when we're young, to interpret these as differences in perspective and, from them, build a three-dimensional view. But even with one eye closed, we don't suddenly switch to interpreting things as if they were in 2-D. Enough clues from perspective, illumination, and shading still arrive via monocular vision to enable us to add depth in our mind's eye. In addition, we can move around or rotate our head to change the angle of sight and add to this other sensory data, such as hearing and touch, to flesh out the 3-D impression. We're so adept at adding a dimension in this way that when we watch a movie on a TV screen, we automatically inject depth, even without the aid of 3-D technology.

The question is, if we have the ability to build 3-D pictures from 2-D input, could we use 3-D visual input to create an impression in our minds of the fourth dimension? Our natural retinas are flat, but electronic technology doesn't have such a limitation. By using enough cameras or other image-gathering devices, stationed in different places, we can collect information from as many directions and perspectives as we like. This alone, however, wouldn't be enough to form the basis of a 4-D view. A genuine four-dimensional observer looking at something in our world would be able to see everything inside a thing simultaneously, in addition to its three-dimensional surface. So, for example, if you had some valuable items locked up in a safe, a 4-D being would see not only all sides of the safe at a single glance but everything inside it as well (and would be able to reach in and take those things if it so chose!). This isn't because the being would have something like X-ray vision that allowed it to see through the walls of the safe, but simply because it had access to an extra dimension. We would similarly have a privileged view of an enclosed space in a 2-D world. Draw a square on a piece of paper, to represent a two-dimensional safe, with some items of jewelry inside it. A flatlander, embedded in the 2-D surface, could see only a view of the outside of his safe—a mere line. We, looking from above the sheet of paper that was his world, would be able to see the lines that formed the walls of the safe and all of its contents at a single glance and could reach in and lift out the 2-D pieces of jewelry. It would mystify the flatlander how the inside of the safe could be observed, or its contents removed, with no gaps in its walls. But in the same way, an observer from the vantage point of a fourth dimension would be able to see all parts, inside and out, of something in 3-D, whether it was a house, a machine, or a human body.

A way to create the illusion of 4-D vision, then, if not 4-D vision itself, would be to have a 3-D retina, consisting of many layers, each layer of which could hold the image of a unique cross-section of a 3-D object. The information from this artificial retina would then be fed directly to a person's brain in such a way that they would have simultaneous access to all of the cross-sections, exactly as a true four-dimensional observer would have. The result would not be an actual 4-D image but something like the view we would have of a 3-D thing if we could look "down" on it from a fourth dimension, which could have some very valuable applications. The first part of the technology required—the 3-D retina—is effectively already available in the form of medical scanners that build up a solid picture of part of the human body from 2-D slices. The second part is at present beyond us, because we don't yet have sufficiently advanced brain-computer interfaces or the neurological knowledge needed to feed into the visual cortex so that the brain can construct an all-perspective, all-at-once image of the thing being observed. However, the dawn of "Human 2.0" may be only a decade or two away. Futurist Ray Kurzweil believes that by the 2030s, we'll be enhancing our brains with nanobots, tiny robotic implants, that connect to cloud-based computer networks. In 2017 technology entrepreneur Elon Musk launched Neuralink, a venture to merge the human brain with artificial intelligence through cortical implants.

As well as putting the technology in place and making the right connections to the brain, to see using a 3-D retina the subject would presumably have to go through a lengthy process of learning how to create mental pictures in this radically new way. However, such an ability could prove invaluable to those involved in areas such as medical diagnosis, surgery, scientific research, and education.

The more difficult step of enabling a person to experience seeing a thing in four dimensions could be done only with simulations, since 4-D objects don't physically exist in our world. Perhaps a computer simulation of a tesseract—the object used by Hinton—would be the simplest place to start. When we look at a 3-D model of a tesseract, we see only one aspect, or projection, of the true four-dimensional shape. Grasping the thing in all its 4-D glory would involve combining multiple projections, seamlessly and simultaneously, in the visual processing parts of our brain. Again, even with all the necessary technology and neural connections in place, it might take a period of training and practice to get the desired effect—to make the fourth dimension, as it were, pop out. But there's no reason in principle that it shouldn't work. By mentally fusing, with the aid of computer technology, a large number of 3-D sections of a 4-D shape, we can hope to know what it is like to see in 4-D.

Mathematics allows us to explore, in depth, what our imaginations alone can't penetrate. It takes us beyond the three dimensions that come naturally to us, so that we can know, in great detail, the properties of things in 4-D and beyond. That allows us to push on with the science we need to do to understand the universe at both the submicroscopic and the cosmic levels. But more, it opens up the possibility of developing the means to visualize dimensions beyond the third for ourselves.

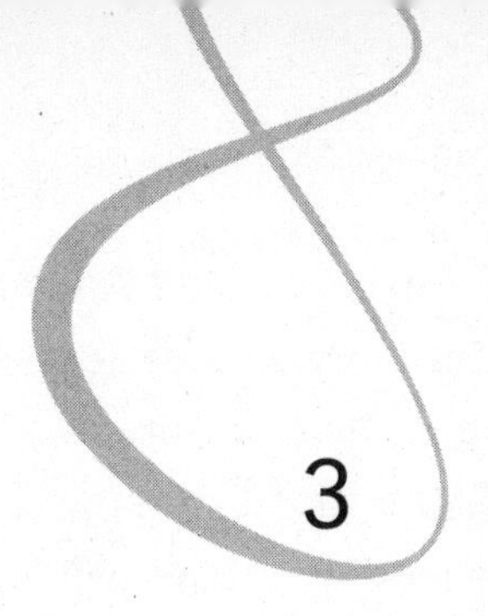

3

CHANCE IS A FINE THING

> So much of life, it seems to me, is determined by pure randomness.
>
> —SIDNEY POITIER

Many things that happen in the world seem utterly unpredictable. We talk about "acts of God," "being in the wrong place at the wrong time," or "pure luck." Serendipity and good or bad fortune seem to dictate so much of what goes on around us. Thanks to math, though, we have a tool to see through this fog of apparent turmoil to make out some order in what otherwise looks like a riot of randomness.

Thoroughly shuffle a deck of cards, and the chances are that you've just done something unique. Almost certainly, no one in the history of the world has ever come up with the deck arranged in that particular order before. The reason's simple: 52 different cards can be arranged in $52 \times 51 \times 50 \times 49 \times \ldots \times 3 \times 2 \times 1$ ways. That's a grand total of about 8×10^{67}, or 80 million trillion trillion trillion trillion trillion, different orderings of the cards. If all the people presently alive were to have shuffled a card deck once every second since the universe began, that would amount to only about 3×10^{27} shuffles, which is an incredibly tiny number by comparison.

Yet there have been claims of decks being shuffled and coming out in exactly the order they started when new. This is actually much more likely than the odds of 1 in 8×10^{67} of getting any other ordering. When first taken out of its wrapper, a card deck has all the suits—hearts, clubs, diamonds, and spades (though not necessarily in that order)—arranged ace, two, three . . . jack, queen, king. If the dealer is so expert as to be able to riffle shuffle without a mistake—splitting the deck in two and exactly interleaving the cards together—the pack can end up back where it was after just eight perfect shuffles. That's why casinos often use a child's approach to shuffling with a brand-new deck, known as "washing the deck," in which the cards are just spread on the table and swished around willy-nilly for a while. Getting a similar level of disorder would take at least seven good but imperfect riffle shuffles. The outcome would then be pretty random; in other words, shown any one card in the deck, the odds of being able to predict the next card, using any fair means available, would be very close to 1 in 51. But would the deck be truly random? What is randomness, and is it *ever* possible to have something that's completely random?

The notion of randomness, or total unpredictability, has been around as long as civilization and probably much longer. Flipped coins and rolled dice most obviously spring to mind as ways we commonly use today to "randomly" decide outcomes. Back in ancient Greece, they tossed *astragali*, or the knucklebones of goats and sheep, in their gambling games. Later they also used regular-shape dice, though where dice first came from isn't known for sure. The Egyptians are thought to have used dice in their game of Senet, five thousand years ago. The Rigveda, a Vedic Sanskrit text dating back to about 1500 BC, also mentions dice, and actual dice games have been found in a Mesopotamian tomb dating back to the twenty-fourth century BC. Greek *tessera* were cubic and had numbers on each side from one to six, but it was only in Roman times that dice like

Pastern bones of an animal, used in games such as knucklebones. (SARAH JOY)

those we use today, in which values on opposite sides add up to seven, first appeared.

It took a long time for randomness to catch the attention of mathematicians. Before that it was mainly thought to be the province of religion. In both Eastern and Western philosophies, the outcome of many events was thought to be in the lap of the gods or some equivalent supernatural force. From China came the I Ching (Classic of Changes), a system of divination rooted in the interpretation of sixty-four different hexagrams. Some Christians based their decision making on the rather simpler method of drawing straws from inside a Bible. Fascinating though these early beliefs were, they had the unfortunate effect of greatly delaying any rational attempts to come to grips with randomness. After all, if eventualities are determined ultimately at some level beyond human comprehension, why bother trying to analyze, logically, why anything happens the way it does? Why try to figure out if there are natural laws that govern the probability of outcomes?

It's hard to believe that those who used *astragali* or dice, in ancient Greek or Roman times, didn't have at least some intuitive feel for the likelihood of certain outcomes. Usually, where money or other material gain is concerned, gamblers and other interested parties quickly catch on to the fine detail of the games they play. So it seems likely that an intuitive appreciation of odds goes back millennia. But the academic study of randomness and probability had to wait until the seventeenth century and the late Renaissance to take off. Spearheading the breakthroughs at this time were French mathematician and philosopher Blaise Pascal, who was also a devout Jansenist, and his compatriot Pierre de Fermat. These two great thinkers tackled a problem that, in simplified form, can be put like this: Suppose two people are playing a coin-tossing game where the first person to get three points wins a pot of money. The game is interrupted with one person leading by two points to one. If the pot is handed out at this stage, what is the fairest allocation? Before Pascal and Fermat, others had thought about this and come up with a variety of possible solutions. Maybe the pot should be divided evenly, since the game was stopped partway and the eventual outcome couldn't be known. But this seemed unfair to the person with two points, who should surely get some credit for being ahead. On the other hand, another suggestion, to hand the whole pot to the person in the lead, looked unfair to the opponent with one point, who would still have had a chance of winning had the game gone on. A third possibility might be to divide the pot based on the number of points gained, so that the player with two points would get two-thirds of the prize and the opponent one-third. On the face of it, this seems fair, but there's a problem with it. Suppose the score was 1–0 at the point the game was interrupted. In this case, if the same rule was applied, the person with one point would receive the entire pot, while the other person, who might still have won if the game ran to its intended conclusion, would get nothing.

Pascal and Fermat found a better solution and, at the same time, opened up a new branch of math. They calculated the probability of each person winning. In order for the person with one point to win, they would have to get two further points in a row, which has a probability of ½ times ½, or ¼. They should, therefore, receive one-quarter of the pot. The rest should go to their opponent. Exactly the same method can be applied to any other problem of this type, although, naturally, the calculations can become more complicated.

In studying this problem, Pascal and Fermat had hit upon a concept known as expected value. In a gambling game, or any situation where chance is involved, the expected value is the average of what you can reasonably hope to gain. For example, suppose you played a game where you rolled a die and won six dollars if you rolled a three. This game has an expected value of one dollar, because there's a one in six chance of rolling a three and one-sixth of the prize money is one dollar. If you played many times, you would earn on average one dollar for each game played. After playing a thousand times, for instance, the average amount you would earn would be one thousand dollars, so if you paid a dollar before playing each time, you would end up breaking even. Notice that even though one dollar is the expected value, it isn't possible to ever win exactly one dollar in this game. It isn't always possible to win the expected value exactly in one game, but if played repeatedly, the expected value is how much you could expect to win on average.

A lottery generally has a negative expected value, so that, from a rational point of view, it's a bad idea to play. (During certain rollovers, depending on the lottery, it may occasionally have a positive expected value.) The same is true of casino games, for an obvious reason: the casino is a business trying to make a profit. Occasionally, though, things can go wrong due to a slight error in calculation. In one instance, a casino changed the payout on just one outcome in blackjack, accidentally making the expected value positive, and lost a fortune within a few hours. Casinos depend on an intimate knowledge of the math of probability theory for their livelihood.

Sometimes coincidences happen that seem so unlikely that people wonder if something funny is going on. A person may win their state lottery twice, or the same numbers may come up in different drawings. Often the media jump on such stories and make a big deal out of their seemingly wild improbability. The truth is, however, that most of us aren't very good at figuring out the likelihood of such events because we start out with some misconceptions. To take the case of someone who won the same lottery twice, it's natural to personalize the problem and think: What are the chances of "me" winning the lottery twice? Obviously, the answer is fantastically small. However, the rare people who win twice tend to have played regularly over a number of years so that any two wins over that period are less remarkable. More important, it has to be borne in mind how many people play the lottery. The vast majority will never win the jackpot once, never mind twice. But with all those people playing, it becomes much less astonishing that someone, somewhere, will take the prize on two occasions.

This misjudgment of odds based on a failure to consider all the possibilities for an event to happen also underlies the so-called birthday paradox, which is really not a paradox at all. With 23 people in a room, the odds that 2 of them will have the same birthday are better than 50–50. It seems that the chances ought to be much smaller than that. You might argue that if it only takes 23 people to find a match, we should all know at least several people who share our birthday, whereas it's always surprising when it happens. But the birthday paradox doesn't ask what are the chances of any 1 person in the room (you, for instance) finding a birthday match, but of any 2 people being born on the same date. In other words, the question is not what are the odds of a particular pair of people sharing a birthday but of *any* pair of people, from all the different possible pairings, being birthday buddies. The odds of this are $1-(365/365 \times 364/365 \times 363/365 \times \ldots \times 343/365) = 0.507$, or 50.7 percent. With sixty people in a group, the odds of a birthday match climb to more than 99 percent. In contrast, for there to be a 50 percent chance of someone having the same birthday as you, 253 people would need to be present.

One reason this might seem unintuitive is that we tend to conflate the two separate questions. Most people don't know 253 people well enough to know their birthdays, so it seems unlikely that someone would randomly share a birthday with them, yet this doesn't mean that it's as unlikely for two other people to share their own birthdays.

Not only ideas in probability can seem counterintuitive, but so too can the notion of randomness. Of the following two sequences of heads (H) and tails (T), which looks the more random?

H, T, H, H, T, H, T, T, H, H, T, T, H, T, H, T, T, H, H, T

or

T, H, T, H, T, T, H, T, T, T, H, T, T, T, T, H, H, T, H, T

Many people might be tempted to say the first because it has an even sprinkling of heads and tails arranged in no obvious pattern. The second sequence has an imbalance of tails and longer runs of the same letter. In fact, one of us (Agnijo) used a random number generator to produce the second, whereas he deliberately constructed the first to look like what a person might come up with if asked to write a random sequence of *H*s and *T*s. A human tends to avoid long runs, deliberately balances the letters, and switches from *H* to *T* and vice versa more often than happens at random.

What about this sequence?

H, T, H, H, H, T, T, H, H, H, T, H, H, H, H, T, H, T, T, T

It may look random, and statistical methods of catching human-produced sequences will conclude that it wasn't made by a person. In reality, it's constructed from the decimal digits of pi (omitting the initial 3), with an *H* for an odd digit and a *T* for an even digit. So are the digits of pi random? Technically, no, because the first decimal digit will always be 1, the second 4, the third 1, and so on, no matter how many times the sequence is generated. If something is fixed and always comes out the same whenever we choose to look at it, it can hardly be random. However, mathematicians do wonder if the decimal digits of pi are statistically random in the sense that they have a uniform distribution: all digits being equally likely, all pairs of digits equally likely, all triplets equally likely, and so on. If they do, then pi is said to be "normal in base 10," which is what the vast majority of mathematicians believe. It's also believed that pi is "absolutely normal," meaning that not only are the decimal digits of pi statistically random, but so too are the binary digits, if pi is written out in the binary number system using just 0s and 1s, the ternary digits, using just 0s, 1s, and 2s, and so on. It's been proved that almost all irrational numbers are absolutely normal, but it turns out to be extremely hard to find a proof for specific cases.

The first example of a known normal number in base 10 was Champernowne's constant, named after English economist and mathematician David Champernowne, who wrote about the significance of it while still an undergraduate at Cambridge. Champernowne invented the number specifically to show that a normal number can and does exist and also how easy it is to construct one. His constant is made up simply of all the consecutive natural numbers: 0.1234567891011121314 . . . and, therefore, contains every possible sequence of numbers in equal proportions. One-tenth of the digits are 1, one-hundredth of pairs of consecutive digits are 12, and so on. Normal in base 10 it may be, but Champernowne's constant is obviously pretty bad at producing sequences that look random—in other words, lacking any kind of discernible pattern or predictability, especially at the start. Nor do we know if it's normal in any other base. Other proven normal constants exist, but like that found by Champernowne, they've been artificially constructed to be normal. It's still to be proven whether pi is normal in any base, let alone absolutely normal.

At the time of writing, the value of pi is known to 22,459,157,718,361, or about 22 trillion decimal digits. Of course, we'll be able to calculate more digits in the future, but those we already know will never change, no matter how many times the calculation is run. The known digits of pi are part of the frozen reality of the mathematical universe, and so they cannot be random. But what about the digits lying beyond those that have been computed? Assuming pi is normal in base 10,

The first couple of hundred digits of pi. (ARTWORK BY JEFF DARLING)

they remain essentially statistically random to us. In other words, if someone asked you for a random string of a thousand digits, it would be a valid response to build a computer to calculate pi to one thousand places more than is presently known and use those places as the random string. Asked for another random string of the same length, you could compute the next (previously unknown) thousand digits. This raises an interesting philosophical question about the nature of mathematical things. To what extent are the decimal places of pi that we haven't yet figured out real? It would be hard to argue that, say, the trillion trillionth digit of pi doesn't exist or that it doesn't have a specific fixed value, even though we don't yet know what it is. But in what sense or form *does* it exist until, at the end of an immensely long calculation, still to be carried out, it pops into a computer's memory?

As a curious aside, it's worth mentioning a discovery made by researchers David Bailey, Peter Borwein, and Simon Plouffe in 1996. They found a fairly simple formula—the sum of an infinite series of terms—for pi that allows any digit of pi to be calculated *without knowing any of the preceding digits*. (Strictly speaking, the digits calculated by the Bailey-Borwein-Plouffe formula are hexadecimal—base-16—digits as opposed to decimal digits.) That seems, at first sight, impossible, and it certainly came as a surprise to other mathematicians. What's more, a computation of, say, the billionth digit of pi, using this method, can be done on an ordinary laptop in less time than it takes to eat a meal at a restaurant. Variations on the Bailey-Borwein-Plouffe formula can be used to find other "irrational" numbers like pi whose decimal extensions go on forever without repeating.

The question of whether anything in pure mathematics is truly random is a valid one. Randomness implies the complete absence of pattern or predictability. Something is unpredictable only if it's unknown and, in addition, there's no basis on which to favor one outcome over any other. Mathematics exists essentially outside of time; in other words, it doesn't change or evolve from one moment to the next. The only thing that does change is our knowledge of it. The physical world, on the other hand, does change, continuously, and often in ways that at first sight seem unpredictable. Tossing a coin is considered to be sufficiently unpredictable that, by common consent, it's taken to be a fair way of making decisions when there are just two possibilities. But whether it can be called random depends on the information available. If, for any given toss, we knew the exact force and angle at which the coin was launched, its rotation rate, the amount of air resistance, and so on, we could (in theory) accurately predict which side would land facing up. The same is true if we drop a slice of buttered toast, except that in this case there's evidence to support the pessimist's view that toast does tend to land butter-side down more than half the time. Experiments have shown that if toast is tossed in the air—surely something that would happen only in a lab or a food fight—the chances of its coming down the messy way are 50 percent. But if the toast is knocked off a table or kitchen counter, or slides off a plate, it will indeed hit the floor butter-side down more often than not. The reason is straightforward: the height from which toast normally gets dropped by accident—waist height or a foot or so on either side—allows the toast just enough time during its fall to make a half turn so that if it starts out, in the conventional way, butter up, it's more likely than not to end up making a grease stain on the floor.

Most physical systems are a lot more complicated than falling toast. And to further complicate the situation, some of them are chaotic, so that little changes or disturbances in the starting conditions may have enormous implications later on. One such system is the weather. Before modern weather forecasting came along, it was anyone's guess what the next day would bring. Meteorological satellites, accurate instruments on the ground, and high-speed computers have revolutionized the accuracy of forecasts, out to about a week or ten days. But beyond that, even the best forecasts, using the finest technology, run into the combined problems of chaos and complexity, including the butterfly effect—the notion that the tiny air current

Hurricane Felix photographed from the International Space Station on September 3, 2007. (NASA)

caused by a butterfly flapping its wings might eventually be amplified so that it becomes a hurricane.

Even with all this complexity, it may seem that no matter what the phenomenon, whether it's the toss of a coin or the global weather system, the same underlying laws of nature are involved and those laws are deterministic. The universe, so it was once believed, is like a giant clockwork mechanism—fantastically elaborate yet ultimately predictable. Two issues, however, stand in the way of this claim. The first harks back to complexity. Even within a deterministic system, one in which the outcome depends on a series of events, each one of which is predictable, knowing the exact preceding state, the whole problem can be so complex that there's no achievable shortcut, allowing us to see in advance what will actually happen. In such systems, the best simulation (for example, run on a computer) cannot outpace the phenomenon itself. This is true of many physical systems but also of purely mathematical ones, such as cellular automata, the most famous example of which is John Conway's Game of Life, which we'll be talking about more in Chapter 5.

The evolution of any given pattern in Life is entirely deterministic yet unpredictable: the outcome becomes known only when every step along the way has been calculated. (Of course, some patterns that do the same thing over and over again, such as oscillating back and forth or moving unchanged after a certain number of steps, are predictable after we know their behavior. But the first time through, we don't know how they're going to behave.) In math things can be unpredictable even if they're not random. But until the turn of the twentieth century, most physicists held the belief that even if we couldn't know every detail of what happens in the physical universe, we could, in principle, know as much as we wanted. If we had enough information, then, using the equations of Newton and Maxwell, we could figure out how events would unfold, to whatever level of accuracy we chose. The dawn of quantum mechanics, however, saw that idea fly out the window.

Uncertainty, it transpires, lies at the heart of the quantum realm: randomness is an unavoidable fact of life in the subatomic world. Nowhere is this capriciousness more evident than in the decay of a radioactive nucleus. True, observations can reveal the half-life of a radioactive substance—the time taken, *on average*, for half of the original nuclei in a sample to break apart. But that's a statistical measure. The half-life of radium 226, for instance, is 1,620 years, so that if we started with one gram of it, we would have to wait 1,620 years for half a gram of the radium to remain, the rest having decayed into radon gas or lead and carbon. Focusing on one individual radium nucleus, though, there's no way to tell if it'll be among the 37 billion nuclei that decay in the next second in one gram of radium 226 or whether it will decay in five thousand years' time. All we can say is that the probability is ½—the same as flipping heads or tails—that it will decay at some point in the next 1,620 years. This unpredictability has nothing to do with shortcomings in our measuring gear or computing power. The randomness, at this fine level of structure, is inherent in the very fabric of reality. As a result, it can affect phenomena, and thereby introduce randomness, on a larger scale. An extreme case of the butterfly effect, for example, would be the decay of a single radium atom influencing the future weather on a large scale.

It may well be that quantum randomness is here to stay. However, there have been physicists, and Einstein was famously one of them, who couldn't stomach the idea (to paraphrase Einstein) that god plays dice with the universe. These opponents of quantum orthodoxy favor the view that, behind the apparent quixotic behavior of things at the ultrasmall level, there are "hidden variables"—factors that determine when particles decay and suchlike, if only we could learn what they are and were able to measure them. If the hidden-variables theory turns out to be true, then the universe would again revert to being nonrandom, and true randomness would exist only as some kind of mathematical ideal. But to date, all the evidence suggests that, on this question of quantum indeterminacy, Einstein got it wrong.

In the looking-glass world of the very small, nothing, it seems, is certain. What we took to be solid little particles—electrons and suchlike—dissolve into waves, and not even material waves but waves of probability. An electron can't be said to be here or there but only more likely to be here than there, its motion and whereabouts governed by a mathematical construct called the wavefunction.

All we are left with is probability, and even that's not an easy concept to pin down. There are different ways of thinking about it. The most familiar is the "frequentist" point of view. In this the probability of an event happening is the limit—the value to which something is heading—of the proportion of times the event occurs. To find out the probability of an event, a frequentist would repeat the experiment many times and see how often the event occurred. For example, if the event occurred 70 percent of the time the experiment was performed, it would have a probability of 70 percent. In the case of an idealized mathematical coin, flipping heads has a probability of exactly ½ because the more the coin is flipped, the closer the proportion of heads approaches the value ½. A real, physical coin doesn't have a probability of exactly ½ of landing heads, for a number of reasons. The aerodynamics of the toss and the fact that, in the case of most coins, the head generally has more mass than the pattern on the other side bias the result slightly. The outcome also depends to some extent on which side is facing up before the toss. The probability is roughly 51 percent that the coin will land the same side up as before tossing, as, during a typical toss, it's marginally more likely to turn an even number of times in the air, but when dealing with mathematical, idealized coins, we can ignore this.

The frequentist approach is to say that the likelihood of something is equal to the long-run chance that it happens. But sometimes, such as for an event that can occur only once, this strategy is useless. An alternative is the Bayesian method, named after eighteenth-century English statistician Thomas Bayes. This bases its calculation of probability on how confident we are in a certain outcome, so that it regards probability as being subjective. For instance, weather forecasters may talk about a "70 percent chance of rain," which essentially means they're 70 percent confident that it will rain. The major difference here between frequentist and Bayesian probability is that the weather forecasters can't simply "repeat" the weather—they need to give a probability of rain on one specific occasion rather than an average probability over many trials. They can use a vast array of data, including what occurred in similar cases, but none of these will be *exactly* identical, so they're forced to use Bayesian probability as opposed to frequentist.

Where differences between the Bayesian and frequentist viewpoints get especially interesting is when they're applied to mathematical concepts. Think about the question of whether the trillion trillionth decimal digit of pi, which is presently unknown, is 5. There's no way in advance of knowing what the answer is, but we do know that once it's been figured out, it won't ever change. We can't repeat a calculation of the digits of pi and get a different answer than the first time it was done. The frequentist viewpoint therefore implies that the probability of the trillion trillionth digit being 5 is either 1 (certainty) or 0 (impossibility)—in other words, it either is or isn't a 5. Suppose pi were to be proven normal, so that we know for certain that every digit has an equal density across the infinite sequence that makes up pi. The Bayesian viewpoint, which is our level of confidence that the trillion trillionth digit is 5, would state that the probability is one in ten, or 0.1 (because if pi is normal, any digit is equally likely to be any number from 0 to 9, until calculated). But the probability after we calculate that far (if we ever do) will then be definitely either 1 or 0. Now, the actual trillion trillionth digit of pi won't change at all, but the probability of its being 5 will change, precisely because we have more information. Information is crucial to the Bayesian viewpoint: more information helps us revise the probability so that it becomes more accurate. Indeed, once we have perfect information (such as by explicitly calculating a digit of pi), frequentist and Bayesian probabilities become equivalent—if we repeat a calculation of a known digit of pi, we know the answer in advance. If we know all details of a physical system (which includes some randomness, such as the decay of radium atoms), we can repeat the exact experiment and get a frequentist probability that exactly matches the Bayesian probability.

While the Bayesian approach may seem subjective, it can be made rigorous in an abstract sense. For example, suppose you had a coin that was biased. It could be biased by any amount from 0 percent heads to 100 percent heads, with each value equally likely. You toss it once, and it comes up heads. It's possible to prove that the probability of a head on the second toss is two out of three using Bayesian probability. However, the initial probability of a head was one out of two, and we didn't change the coin. The Bayesian viewpoint says that while the first head will not directly affect the probability of the second head, it gives you more information about the coin that allows you to refine your estimate. A coin heavily biased toward tails is highly unlikely to flip heads, and a coin heavily biased toward heads is much more likely to flip heads.

Taking a Bayesian approach also helps avoid a type of paradox first pointed out by German logician Carl Hempel in the 1940s. When people see the same principle, such as the law of gravity, operating without fail over a long period of time, they naturally assume that it's true with a very high probability. This is inductive reasoning, which can be summed up as follows: if things are observed that are consistent with a theory, then the probability that the theory is true increases. Hempel, however, pointed out a snag with induction, using ravens as an example.

All ravens are black, so the theory goes. Every time a raven is seen to be black and no other color—ignoring the fact that there are albino ravens!—our confidence in the theory "All ravens are black" is boosted. Here, though, is the rub. The statement "All ravens are black" is logically equivalent to the statement "All nonblack things are nonravens." So, if we see a yellow banana, which is a nonblack thing and also a nonraven, it should bolster our belief that all ravens are black. To get around this highly counterintuitive result, some philosophers have argued that we shouldn't treat the two sides of the argument as having the same strength. In other words, yellow bananas should make us believe more in the theory that all nonblack things are nonravens (first statement), without influencing the belief that all ravens are black (second statement). This seems to fit with common sense—a banana is a nonraven, so observing one can tell us about nonravens but tells nothing about ravens. But it's a suggestion that's been criticized on the basis that you can't have a different degree of belief in two statements that are logically equivalent, if it's clear that either both are true or both are false. Maybe our intuition in this matter is at fault, and seeing another yellow banana really *should* increase the probability that all ravens are black. Adopting a Bayesian stance, however, the paradox never arises. According to Bayes, the probability of a hypothesis *H* must be multiplied by this ratio:

$$\frac{\textit{Probability of observing X if H is true}}{\textit{Probability of observing X}}$$

where *X* is a nonblack object that's a nonraven and *H* is the hypothesis that all ravens are black.

If you ask someone to select a banana at random and show it to you, then the probability of seeing a yellow banana doesn't depend on the colors of ravens. You already know beforehand that you'll see a nonraven. The numerator (the number on top) will equal the denominator (the number on the bottom), the ratio will equal one, and the probability will remain unchanged. Seeing a yellow banana won't affect your belief about whether all ravens are black. If you ask someone to select a nonblack thing at random and they show you a yellow banana, then the numerator will be bigger than the denominator by a tiny amount. Seeing the yellow banana will only slightly increase your belief that all ravens are black. You would have to see almost every nonblack thing in the universe and see that they were all nonravens before your belief in "All ravens are black" went up significantly. In both cases, the result agrees with intuition.

It may seem odd that information is connected to randomness, but in fact the two are closely related. Imagine a string of digits made only of 1s and 0s. The string 1111111111 is completely orderly and, because of this, contains practically no information (only "repeat 1 ten times"), just as a blank canvas where every point is white tells us almost nothing. On the other hand, the string 0001100110, which was generated randomly, has the maximum amount of information possible for its length. The reason for this is that one way of quantifying information is the amount by which the data can be compressed. A truly random string can't be written in any shorter way while retaining all of its information. But a long, constant string with only 1s, for example, can be compressed enormously just by listing the number of 1s in the string. Information and disorder are intimately related. The more disordered and random a string is, the more information it has within it.

Another way to think of this is that in the case of a random string, revealing the next bit gives the maximum amount of information possible. On the other hand, if we see the string 1111111111, it is trivial to guess the next bit. (This only applies to one string as a whole, not part of another string. An arbitrarily long random string will contain 1111111111 infinitely often.) Useful stimuli, as far as we're concerned, must necessarily occupy a middle ground between these extremes of information. For example, a photograph with minimum information would be a blank monochromatic picture, and a book would be a long repetition of pages filled with one letter. Neither of these is in any way interesting in terms of its information content. However, a photograph with maximum information would look like a random mess of static, and a book would be a jumble of random letters. These again would not appeal to us. What we need, and what is most useful to us, is something in between. A conventional photo conveys information, but in a form and quantity that we can understand. If one pixel is one color, pixels immediately adjacent to it are likely to be very similar. We know this and can use it to compress pictures without losing the information. The book you're reading right now is mostly just a string of letters and spaces, with punctuation marks. Unlike in extreme books that contain a jumble of symbols or all the same one, these letters fall into structured patterns known as words, some of which occur occasionally and some, such as "the," that recur extremely frequently. In addition, these words follow certain rules, known as grammar, to form sentences and so on, so that ultimately the reader can understand the information being conveyed. This simply doesn't happen in a random hodgepodge.

In his short story "The Library of Babel," Argentine writer Jorge Luis Borges tells of a library, vast—possibly infinite—in size, that contains a dizzying number of books. All the books are of identical format: "Each book contains four hundred ten pages; each page, forty lines; each line approximately eighty black letters." Only twenty-two alphabetic characters plus a comma, period, and space are used throughout, but every possible combination of these characters that follow the common format occurs in some book in the library. Most books appear to be just a meaningless jumble of characters; others are quite orderly but still devoid of any apparent meaning. For example, one book contains just the letter *M* repeated over and over. Another is exactly the same except that the second letter is replaced by an *N*. Others have words, sentences, and whole paragraphs that are grammatically correct in some language but are nevertheless illogical. Some are true histories. Some purport to be true histories but are, in fact, fictional. Some contain descriptions of devices yet to be invented or discoveries yet to be made. Somewhere in the library is a book that contains every combination of the basic twenty-five symbols that can be imagined or written down in the given format. Yet, of course, it's all useless because without knowing in advance what's true or false, fact or fiction, meaningful or meaningless, such exhaustive combinations of symbols have no value. It's the same with the old idea of monkeys randomly hitting the keys of typewriters and eventually, given enough time, coming up with the works of Shakespeare. They would also come up with the solutions to every major problem in science (after countless trillions of years). The trouble is that they would also come up with every nonsolution and every convincing refutation of the true solutions and, for the most part, mind-numbing quantities of pure gobbledygook. Having the answer before you is no use at all if you also have every other possible variant of the symbols that make up the answer and you have no way of knowing which one is right.

In a sense, the World Wide Web, with its vast collection of knowledge available alongside an even more enormous body of gossip, half-truths, and pure gibberish, is becoming like Borges's library—a repository of everything from the profound to the nonsensical. There are even websites that simulate the Library of Babel, generating, in an instant, pages of random strings of letters, which may or may not include real words or meaningful scraps of information. When there are so much data available at our fingertips, who or what can be trusted to be the arbiters of reason and fact? Ultimately, because the information exists as a collection of digits, inside electronic processors and memories, the answer must lie with mathematics.

In the more immediate future, mathematicians are working to develop an overarching theory of randomness that might connect seemingly very different phenomena in science, from Brownian motion to string theory. Two researchers, Scott Sheffield at the Massachusetts Institute of Technology (MIT) and Jason Miller at the University of Cambridge, have found that many of the 2-D shapes or paths that can be generated by random processes fall into distinct families, each with its own sets of characteristics. Their classification has led to the discovery of unexpected links between what, on the face of it, look like totally disparate random objects.

The first kind of random shape to be explored mathematically is the so-called random walk. Imagine a drunkard who starts from a lamppost and staggers from one point to another, each step (assumed to be of equal length) being taken in a random direction. The problem is to work out how far from the lamppost he's likely to be after a given number of steps. This can be reduced to a one-dimensional case—in other words, just movements back and forth along a line—by supposing that at each step, a coin is tossed to decide which way to move, right or left. The problem was first given a real-world application in 1827 when English botanist Robert Brown drew attention to what became known as Brownian motion—the haphazard jiggling of pollen grains in water when looked at through a microscope. Later, it was realized that the jiggling was due to individual water molecules striking the pollen grains from different, random directions, so that each pollen grain behaved like the drunkard in our original example. It took until the 1920s for the mathematics of Brownian motion to be fully worked out, by American mathematician and philosopher Norbert Wiener. The trick is to figure out what happens to the random-walk problem as the steps and the time between them are made smaller and smaller. The resulting random paths look very much like those of Brownian motion.

More recently, physicists have become interested in random motion of a different kind, involving not particles following 1-D curves but incredibly tiny, wriggling "strings" whose motion can be represented as 2-D surfaces. These are the strings of string theory, a leading but as yet unproven theory of the fundamental particles that make up all matter. As Scott Sheffield put it: "To make sense of quantum physics for strings, you want to have something like Brownian motion for surfaces." The beginnings of such a theory came in the 1980s, thanks to physicist Alexander Polyakov, now at Princeton University. He came up with a way of describing these surfaces that's now known as Liouville quantum gravity (LQG). A separate development, called the Brownian model, also described random 2-D surfaces but gave different, complementary, information about them. Sheffield and Miller's big breakthrough was to show that these two theoretical approaches, LQG and the Brownian model, are equivalent. There's still work to be done before the theory can be applied directly to problems in physics, but eventually it may prove to be a powerful unifying principle that operates on many scales, from the fantastically small scale of strings to the everyday level of phenomena such as the growth of snowflakes or mineral deposits. What's already clear is that randomness lies at the heart of the physical universe, and at the heart of randomness is math.

Something that's truly random is unpredictable. There's no way of telling what the next member of a random sequence will be. In physics there's no way of telling when a random event, like the decay of a radioactive nucleus, will take place. If something is random, it's said to be nondeterministic because we can't figure out, even in principle, what comes next from what's already happened. In everyday speech, we often say that if something is random, it's chaotic. "Randomness" and "chaos" are used in ordinary language almost interchangeably. But in math there's a big difference between the two, a difference we can appreciate by venturing into the strange realm of fractional dimensions.

4

PATTERNS AT THE BRINK OF CHAOS

> Mathematics has beauty and romance. It's not a boring place to be, the mathematical world. It's an extraordinary place; it's worth spending time there.
>
> —MARCUS DU SAUTOY

Look up "chaos" in a thesaurus, and you'll find synonyms like "turmoil," "lawlessness," and "anarchy." But the kind of chaos with which mathematicians and scientists deal, in a relatively new field known as chaos theory, is very different. Far from being a disorderly free-for-all, it follows rules, its onset can be foretold, and its behavior is revealed in patterns of exquisite beauty. Digital communications, modeling of the electrochemistry of nerve cells, and fluid dynamics are among the practical applications of chaos theory. But we'll take a more scenic approach to the subject by way of a disarmingly simple question.

"How Long Is the Coast of Britain?" That's part of the title of a paper by Polish-born French American mathematician Benoît Mandelbrot, a theorist at the IBM Thomas J. Watson Research Center, published in the journal *Science* in 1967. It seems a straightforward-enough problem to solve. Surely, all you would need to do is measure accurately all the way around the coast, and that would be it. In fact, though, the length you measure depends on the scale you use, but in such a way that the length can increase without bound (as opposed to converging to a fixed value)—at least down to the atomic scale. This puzzling result, that the coastline of an island, or a country or a continent, doesn't have a well-defined length, was first talked about by English mathematician and physicist Lewis Fry Richardson several years before Mandelbrot expanded on the idea.

Richardson, a pacifist interested in the theoretical roots of international conflict, wanted to know whether there was a connection between the chances of two countries going to war and the length of their common border. While researching this problem, he found marked discrepancies between the values quoted by different sources. The length of the Spanish-Portuguese border, for instance, was once said by Spanish authorities to be just 987 kilometers (613 miles) long, while the Portuguese gave it as 1,214 kilometers (754 miles). Richardson realized that this kind of spread of measurements came about not because anyone was necessarily wrong but because they were using different "yardsticks," or minimum-length units, in their calculations. Measure the distance between two points on a wiggly coastline or border with a ruler that's 100 kilometers long, and you'll get a smaller value than using a ruler that's half the length. The shorter the ruler, the smaller the wiggles it can take into account and

Great Britain and Ireland photographed by NASA's Terra satellite on March 26, 2012. (NASA)

include in the final answer. Richardson showed that the measured length of a wiggly coast or border increased without limit as the yardstick, or unit of measurement, shrank and shrank. Evidently, in the case of the Spain-Portugal border, the Portuguese had done their measuring with a shorter-length unit.

At the time he published his findings, in 1961, no one paid much attention to this surprising discovery—what's now called the Richardson effect or coastline paradox. But looking back, it's seen as an important contribution to the development of an extraordinary new branch of mathematics that Mandelbrot, the man who made it famous, eventually described as "beautiful, damn hard, increasingly unuseful." In 1975 Mandelbrot also coined a name for the weird things at the heart of this novel field of research: *fractals*. A fractal is something, such as a curve or a space, with a fractional dimension.

To be a fractal, all a shape needs is to have a complex structure at all scales, no matter how small. The vast majority of curves or geometrical figures we come across in math aren't fractals. A circle, for instance, isn't a fractal because if we progressively zoom in on part of the circle's circumference, it looks more and more like a straight line, after which nothing new is revealed at higher magnification. A square isn't a fractal, either. It retains the same structure at its four corners, and everywhere else looks just like a straight line after zooming in. To qualify as a fractal, it isn't even enough to have complex structure at one, or finitely many, point(s); there has to be that kind of structure at all points. The same is true of shapes in three or more dimensions. Spheres and cubes, for instance, aren't fractals. But there are many shapes, in many different dimensions, that are.

Going back to the coastline of Great Britain, a small-scale map shows only the largest bays, inlets, and peninsulas. Go to a beach, however, and you'll see much smaller features—coves, headlands, and so forth. Look even closer, with a magnifying glass or microscope, and you would be able to discern the minuscule structure of the edges of every rock on the shore, down to smaller and smaller scales. In the real world, there's a limit as to how far it's possible to zoom in. Below the level of atoms and molecules—and perhaps even well before then—it becomes meaningless to talk about more detail to do with the length of coastlines, and, in any case, the length changes, minute by minute, due to erosion and the ebb and flow of the tides. Nevertheless, the coast of Great Britain and the outline of other islands and countries are pretty good approximations to fractals, and this explains why different sources of data give different values for the lengths of borders. Looking at a map of the whole of Great Britain, you wouldn't be aware of all the little indentations that you would see if you actually walked around the coast and, so based on the map, would come up with a much shorter length. Simply strolling along the beach, you would miss all the fine structures of the rocks and get a much shorter length than if you used a foot-long ruler or something even more precise to measure all the little ins and outs at a small scale. The length actually grows exponentially, the more precise the measurement gets, rather than approaching, ever more closely, a final "true" figure. In other words, if you use measuring equipment with enough resolution, the length you obtain can be as great as you like (again, within the limits set by the atomic nature of matter).

As well as natural fractals, such as coastlines, there are many purely mathematical fractals. A simple way to make one is to divide a line into three equal sections; draw an equilateral triangle, pointing outward, that has the middle section as its base; and then remove the section that is the base. This process is then repeated, starting from each of the four resulting line sections, and repeated again for each of the new shorter sections, and so on, for as long as you like, or forever. The final result is known as the Koch curve, after Swedish mathematician Helge

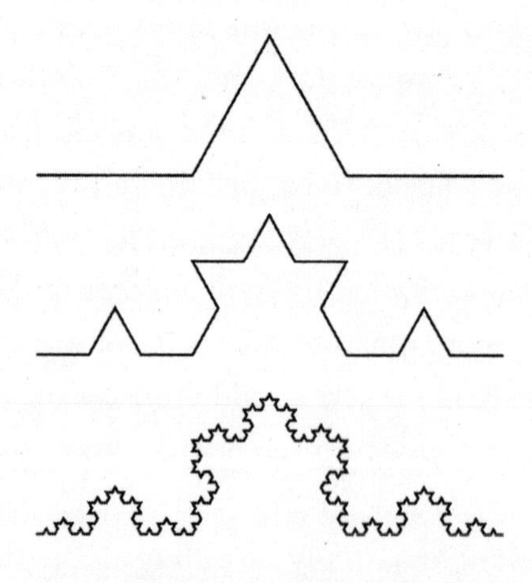

The first, second, and fourth iterations of the Koch curve. (AGNIJO BANERJEE)

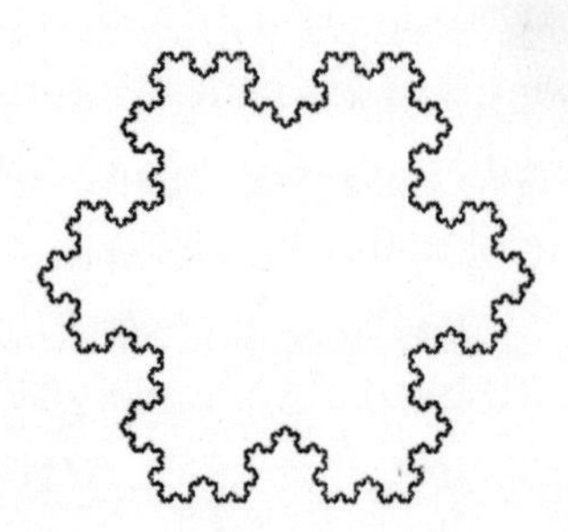

The Koch snowflake. (AGNIJO BANERJEE)

von Koch, who wrote about it in a paper in 1904. Three of these Koch curves can be joined to form what's known as the Koch snowflake. The Koch curve was one of the earliest fractal shapes to be constructed. A couple of other now familiar fractals were described mathematically in the first quarter of the twentieth century by Polish mathematician Wacław Sierpiński: the Sierpiński sieve (or gasket) and carpet. To make the sieve, Sierpiński started with an equilateral triangle and divided it into four, each half the size of the original. Then he removed the central one to leave three equilateral triangles, repeated the process with each new triangle, and kept on doing this, over and over again. Although such objects have been the subject of serious mathematical study for not much more than a century, artists have known about them since antiquity. The Sierpiński sieve, for instance, appeared in Italian art, such as mosaics in the cathedral of Anagni, going back to the thirteenth century.

Among the most interesting and counterintuitive properties of fractals is their dimension. On hearing the word "dimension," a couple of ideas generally spring to mind, one to do with the overall size of something (as in the "dimensions" of a room) and the other that refers to a unique spatial direction, which is the kind of dimension we talked about in Chapter 2. We say that a cube is three-dimensional because it has sides that lie in three different directions at right angles to one another. This second intuitive understanding of dimension, which relates to the number of perpendicular directions it's possible to travel in, is roughly equivalent to what in mathematics is called the topological dimension. A sphere has topological dimension 2 because we can travel along it in the directions given by North and South or by East and West. A ball, on the other hand, has topological dimension 3 because it also has an up and down, where down is toward the center and up is away from the center, as on earth. The topological dimension can even be 4 or greater, as we saw in Chapter 2 (for example, a tesseract has topological dimension 4), but is always a whole number. However, the fractal dimension is different and measures, roughly speaking, how well a curve fills the plane or how well a surface fills space.

There are many different forms of the fractal dimension, but one of the easiest to grasp is the box-counting dimension, also known as the Minkowski-Bouligand dimension. To calculate it in the case of the coastline of Great Britain, we would cover a map of the coastline with a transparent grid of small squares and count the number of boxes that overlap the coast. Then we would halve the size of the boxes and count again. If this was done with a straight line, the number of boxes would simply double, or go up by a factor of 2^1, where the exponent (1) is the box-counting dimension. If it was done with a square, the number of boxes would quadruple, or go up by a factor of 2^2, giving a dimension of 2, and in the case of a cube (using a three-dimensional grid), the number of boxes would increase by a factor of 8, $= 2^3$, because a cube is three-dimensional.

Most of the shapes we're used to thinking about have a whole-number dimension—1, 2, or 3. But fractals are different. In the case of the Koch snowflake, to simplify things we can use the fact that each element of it, the Koch curve, is made of four smaller Koch curves. If we reduce the size of the boxes in our grid by a factor of 3, we can split the Koch curve into four smaller copies, each one-third the size. Each smaller copy has as many small boxes overlapping with it as the larger copy had with the original boxes, so the total number of boxes has been multiplied by 4. This gives us the dimension, *d*, of the Koch curve (and also the Koch snowflake, which is built of Koch curves) from the relationship $3^d = 4$. Solving this equation gives a value for *d* of about 1.26, so a Koch snowflake is 1.26-dimensional. Think of this as telling us how much more wiggly a Koch snowflake is than a straight line at any scale we care to choose. Or think of it another way, as the extent to which the Koch snowflake fills the (2-D) plane in which it lies; it's too detailed to be one-dimensional but too simple to be two-dimensional. A line goes no way at all toward filling a plane because not only is it infinitely thin but also it's simple in form. A fractal, like a Koch curve, is infinitely thin but so convoluted that between any two points, no matter how close they may appear when we zoom out, there's an infinitely long distance along the curve.

Applying the box-counting method to the Sierpiński sieve, we end up with a value for *d* of 1.58. This state of affairs, in which objects can have a noninteger dimension, seems very strange. And the strangeness spills over from the realm of the purely mathematical to things in the physical world.

Fractals such as the Koch snowflake and Sierpiński sieve are self-similar, which means that they're made up of successively smaller copies of themselves. In nature most fractals aren't exactly self-similar. However, they're statistically self-similar, and so we can still work out their fractal dimension by applying the box method as before. When this is done, the fractal dimension of the coastline of Great Britain turns out to be about 1.25, remarkably similar to that of the

Koch snowflake. In simple language, what this means is that Britain's coast is one and a quarter times more wiggly, or rough, at all scales than a straight line or any other simple curve. South Africa, by comparison, has a much smoother coastline and a correspondingly lower fractal dimension of 1.05. Norway, with its impressive number of deep and convoluted fjords, scores a fractal dimension of 1.52. The concept can be applied to other natural fractals, one notable example being the human lung. Because the lung itself is obviously three-dimensional, you might expect its surface to be two-dimensional. However, the lung has evolved to have an enormous surface area—between eighty and one hundred square meters, or roughly half the area of a tennis court—in order to be able to exchange gases as quickly as possible. So convoluted is the lung's surface, with all its countless folds and tiny air sacs, or alveoli, that it almost fills the space it contains. Its box dimension works out to be about 2.97 so that, measured in this way, it's almost three-dimensional.

In the real world, there are only three spatial dimensions, but time is also sometimes considered to be the "fourth dimension." It's no surprise, then, that fractals can exist in time as well as in space. An economic example is the stock market. Over time, there may be large upward and downward fluctuations in the value of stocks, some of which take place over a period of years and others (such as crashes) that can happen very quickly. As well as this, there are smaller fluctuations, when stocks rise and fall seemingly independently of the large-scale trends, and also tinier fluctuations that happen many times a day, as individual stocks rise and fall by slight amounts. With the computerization of the stock market, these trends can be followed down to very small slices of time, from minute to minute, and even from one second to the next.

Another example of a time-based fractal is something we have already come across—the changing length of the coastline of an island, such as Great Britain. At any given moment, the coastline is a purely spatial fractal, the measured length of which depends on the magnification factor. But over time, as mentioned earlier, there are additional variations because of continual erosion (and deposition), the coming and going of tides and even of individual waves, and the almost imperceptible rise or fall of whole landmasses due to tectonic activity.

Of all fractals known to mathematicians, one stands out because of its incredible intricacy. Not only does this fantastic shape have structure at all scales, but at different points at different scales, it can look like two completely different fractals! It's the famous Mandelbrot set, which was described by American author James Gleick in his book *Chaos*, perhaps questionably, as "the most complex object in mathematics." Although it carries Benoît Mandelbrot's name, there has been some dispute over who actually discovered it. Two mathematicians have argued that they had found it independently at about the same time, while another, John Hubbard of Cornell University, has pointed out that, in early 1979, he went to IBM and showed Mandelbrot how to program a computer to plot out parts of what, after Mandelbrot's publication of a paper on the object, the following year, became known as the Mandelbrot set. The feeling is that Mandelbrot was a good popularizer of the field of fractals and devised clever ways to display fractal images but that he was less than generous in giving credit to other mathematicians where credit was due.

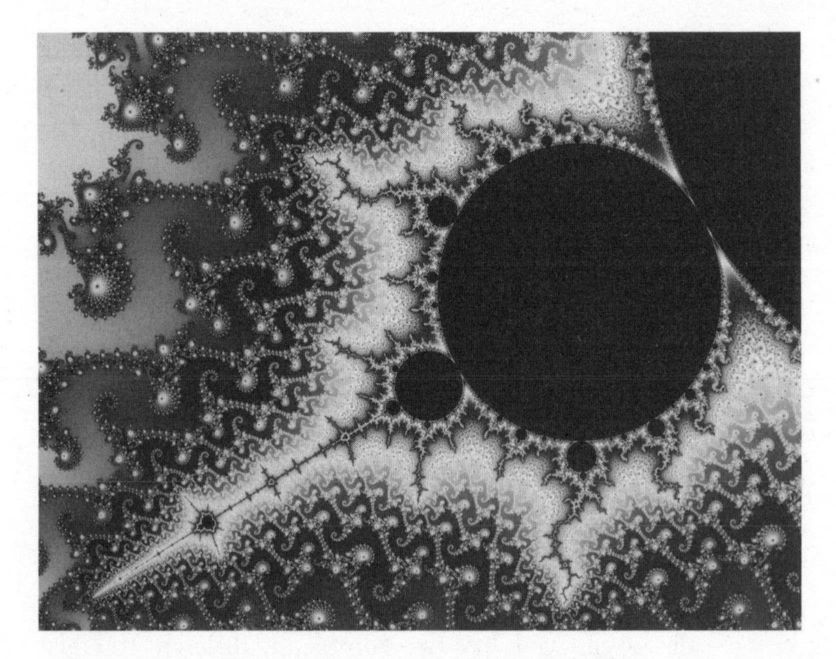

A partial view of the Mandelbrot set. (WOLFGANG BEYER)

Fabulously labyrinthine though the Mandelbrot set is, it arises from a very simple rule, which is just applied over and over again. In essence, the rule is this: take a number, square it, and add to it a fixed number. Then feed the result back into the formula, and keep going around and around, or iterating, in this way. The numbers in question are complex numbers—"complex" meaning that each is made up of a real number part and an "imaginary" one (a number times the square root of minus one). The fractal shape emerges when the real and imaginary values of each number are plotted on a graph.

To elaborate on this a little, say we start with a complex number (z) and a constant (c), which is also a complex number. Having chosen a value for z, we apply to it the rule "Multiply z by itself and add c," or $z^2 + c$. This gives us a new value for z, which we then feed back into the same rule to obtain the next z value. Some values of z will stay the same, and others will repeat in a cycle before eventually returning to their original value. Any of these values, which either stays the same or repeats in a cycle, is said to be stable if we can change z very slightly and have the new value follow a path that stays very close to the original path. This is like the situation of a ball in a valley. If the ball is moved slightly, it will just roll back to its original position and is therefore stable. A ball on the peak of a mountain, on the other hand, even if nudged slightly, will roll down the mountain and follow a completely different path, so that this position at the mountain peak is unstable.

The stable points, out of those that stay the same or are in a cycle, are known as attractors. There are also other points, not necessarily very close to an attractor to start with, that get closer and closer to it as the iteration process continues. These form the "basin of attraction" for c. Other points may get farther and farther away, diverging to infinity. The boundary of the basin of attraction is known as the Julia set for c. Julia sets are named after French mathematician Gaston Julia, who, along with his compatriot Pierre Fatou, did pioneering work on the subject of complex dynamics in the early 1900s. If you iterate any point on the Julia set, the resulting point will stay on the Julia set, but may move around it without settling into a repeating pattern.

The simplest possible Julia set is when $c = 0$ because then the rule for getting new values of z becomes simply "Multiply z by itself." What happens to a complex number (z) when it's iterated in this way? If z starts off inside the unit circle (a circle of radius 1) centered on 0, it will rapidly spiral in toward 0. If z is outside this circle, it will rapidly spiral out to infinity. The Julia set is therefore the boundary of the unit circle, the basin of attraction is everywhere inside the unit circle, and the attractor is the point 0. Imagine the Julia set with $c = 0$ to be like a steel ball placed exactly between two magnets; the ball will stay put (on the Julia set, although in practice z can move about unpredictably as long as it stays on the Julia set), but if it were placed even slightly differently, it would rapidly be attracted to a magnet. In this case, one of the magnets represents 0 and the other infinity.

This Julia set isn't terribly interesting and certainly isn't a fractal. However, apart from $c = 0$, Julia sets do indeed form fractals and may come in many different shapes. Sometimes a Julia set is connected, and sometimes it isn't. When it isn't, it takes the form of Fatou dust, which, as the name suggests, is a cloud of disconnected points. Fatou dust is actually a fractal with a dimension less than 1.

The Mandelbrot set is the set of all values of c for which the Julia set is connected. It's one of the most recognizable yet counterintuitive fractals. Although the Mandelbrot set is connected, there are tiny specks that don't seem to be joined to it at all but in fact are, by means of extremely slender filaments. When magnified these specks are found to be replicas of the entire Mandelbrot set, which may seem surprising at first but actually fits in with our understanding of the nature of fractals. These offshoots are imperfect replicas, yet no two of them are exactly alike—and for a very good reason that turns out to be one of the most profound facts about the Mandelbrot set. If you zoom in on any point on the boundary of the Mandelbrot set, it begins to look more and more similar to the Julia set at that point. The Mandelbrot set, a single fractal, contains infinitely many completely different fractals in the form of a vast array of Julia sets, all along its boundary. Indeed, the Mandelbrot set has been called a catalog of Julia sets. Its boundary is so extraordinarily complex that it turns out to be two-dimensional, though it's conjectured to have zero area.

Fractals often exemplify a straightforward yet counterintuitive principle: it's possible to generate fantastically complex structures and patterns from extremely simple rules. The Koch snowflake is conjured up by a rule that a child could understand (just add an equilateral triangle onto the middle third of each line) yet has an elaborate, albeit regular, structure. The Mandelbrot set is vastly more complex but, again, springs from a disarmingly simple recipe of instructions. You start with the function $z^2 + c$ and, by examining properties and asking questions, arrive at a fractal that has bewildering complexity, looking completely different at different points. Using a computer as a microscope, it is possible to zoom in on any part of the Mandelbrot set and discover pattern within pattern, never exactly repeating and never reaching an end.

Fractals have one other interesting property. The fractal dimension of the Koch snowflake, as we've seen, is 1.26, which gives an idea of how "rough" it is, or how well it fills the plane. If we take an arbitrary line that intersects the Koch snowflake, the intersection is almost always itself a fractal with dimension 0.26. (There are a few degenerate cases, such as a line of symmetry, which results in two isolated points for a fractal dimension of 0.) This is true for any fractal with dimension between 1 and 2 inclusive. For example, almost all lines that intersect the boundary of the Mandelbrot set form a fractal with dimension 1, though they consist of disconnected points and have length 0.

If we consider the same with fractals of dimension less than 1, something else happens. These fractals all consist of a cloud of isolated points. An example is Fatou dust. The surprising result is that almost all lines that intersect Fatou dust do so at only one single point, for a fractal dimension of 0, while almost all lines in general, even if restricted to those passing through the Fatou dust, will never intersect it.

These fractals all exist in two-dimensional space. We can even go down to one-dimensional space and find fractals, consisting of disconnected clouds of points and having a fractal dimension of 1 or less. The most famous example is the Cantor set. Start by taking a line segment. Remove the middle third, leaving two separate line segments. Do the same over and over again. In the end, all line segments have been reduced to disconnected points that constitute a fractal with a fractal dimension of approximately 0.63.

Fractals are related to another phenomenon in mathematics, known as chaos. Both arise from iterated functions, or rules that are cycled through over and over. Each iteration takes the state of the previous iteration as an input to produce the next state. In the case of fractals, the iterations generate a repeating or somewhat repeating pattern to which there's no end no matter how much we zoom in. The distinguishing features of chaos are complexity that lacks any repeating pattern and an extreme sensitivity to initial conditions, or the starting state of the system.

The word "chaos" itself is Greek and in its original form means "void" or "emptiness." In classical and mythological notions of creation, it was the formless state out of which the universe emerged. In math and physics, chaos or a chaotic state is equivalent to randomness or lack of pattern. But chaos theory is different from all these and refers to the behavior of nonlinear dynamic systems under certain conditions. The behavior of the weather gives a familiar example. Nowadays we can easily forecast the weather in the short term, over a few days or a week, and get it right much of the time. But there are no reliable forecasts for longer timescales, such as a month. That is because of chaos.

Suppose we take the weather, starting from a particular initial condition. We can compute the forecast into the future from those conditions. However, if we change the conditions at the start by even a minuscule amount, the weather will very soon become unrecognizably different. This fact is what led to the discovery of chaos in the first place, by American mathematician and meteorologist Edward Lorenz. In the 1950s, he was working on a mathematically simplified model of the weather. He plugged numbers into his computer and generated a graph but was interrupted in midcomputation and had to restart the program. Instead of going back to the very start (which would have taken too much time), he started at a point in the middle and input the results from there. The graph he got at first seemed to agree with his previous one but soon rapidly diverged, as if it were a completely different graph. The reason was that a computer stores a few more digits than it outputs for rounding purposes. When Lorenz restarted the program, those digits were lost, so the input was imperceptibly different from the initial result at that point. The difference was amplified by the program until it diverged rapidly. This gave rise to a principle that Lorenz called the butterfly effect: a reference to the fact that if a butterfly flaps its wings today, it might lead to a hurricane a month later.

Simpler equations than those used to predict the weather can show this same effect, revealing the point at which pattern and predictability break down and chaos takes over. Let's say we start off with some value of x, where x can take any value between and including 0 and 1. Then we multiply x by $(1-x)$ and also by a constant number k, where k is between 1 and 4 inclusive. The new value of x is cycled back into this formula, again and again. In mathematical jargon, the process can be summarized as: $x \rightarrow kx(1-x)$ for $0 \leq x \leq 1$ and $1 \leq k \leq 4$. What we find is that for values of k that are less than or equal to 3, there's an attractor consisting of a single point, with every value of x (apart from 0 and 1) converging to it. For values of k between 3 and 3.45, the attractor consists of two points, which alternate. When k lies between 3.45 and 3.54, the attractor consists of four points, then eight, and so on, doubling more and more often. At approximately $k = 3.57$, a big change takes place, and the doubling goes from happening faster and faster to happening an infinite number of times. At this point the system can never settle down to a steady pattern and becomes completely chaotic. Chaos emerges when a predictable system becomes completely unpredictable. For example, in this case, when k is less than 3, it's simple to predict that after, say, one hundred iterations, a point will be very close to the single attractor. For k greater than 3.57, there's no way for us to predict the long-term behavior of any point.

The doubling of attractor points, from one point to two to four, and so on, which happened when k exceeded the value 3 in the example we just looked at, is governed by an important mathematical constant known as the Feigenbaum constant. We can see how this important number emerges in the lead-up to chaos. The first phase, with a cycle of one point, has length 2, because it lasts from $k = 1$ to $k = 3$. The second, with a cycle of two points, has length approximately 0.45 because it lasts from $k = 3$ to $k = 3.45$. The ratio 2/0.45 is approximately 4.45. The third phase has length approximately 0.095. The ratio 0.45/0.095 is approximately 4.74, and so on. These ratios eventually converge to the Feigenbaum constant, about 4.669. Each phase lasts exponentially shorter than the last, so that by $k = 3.57$ the cycling has occurred infinitely many times.

The Feigenbaum constant emerges from the process we have just considered, but what makes it fundamental to chaos theory is that it can be found in all similar chaotic systems. No matter what the equation, as long as it satisfies some basic conditions, it will have cycles that double in length according to the Feigenbaum constant.

To see how chaotic processes can generate fractals, we could take the iterative process above and plot the attractors for each k. Most of what appears after $k = 3.57$ is pure chaos, but there are a few values of k for which there's a finite attractor. These are known as islands of stability. One such island occurs around $k = 3.82$, where we find an attractor consisting of just three values. Zoom in on any one of these values, and what we see looks similar, though not exactly identical, to the entire graph.

In his pioneering studies of chaos, Lorenz also found a new kind of fractal, known as a strange attractor. Ordinary attractors are simple in the sense that points converge to them and then follow a fixed cycle along them. But strange attractors behave differently, as we'll see. Lorenz used a system of differential equations to form the first. When he zoomed in on any point on it, it gave the appearance of infinitely many parallel lines. Any point on the attractor followed a chaotic path along the attractor, never returning exactly to its original position, and two points that started very close to each other rapidly diverged and ended up following very different paths. For a physical analogy of this, imagine a Ping-Pong ball and an ocean. If the ball is released above the ocean, it will rapidly fall until it reaches the water. If it's released below the surface, it will rapidly float up. But once it's on the ocean's surface, its motion becomes unpredictable and chaotic. Likewise, if a point is not on a strange attractor, it will rapidly move toward it. Once it's on the strange attractor, though, it moves around it in a chaotic manner.

Fractals are fascinating to explore and among the most visually stunning objects in math. But they're also profoundly important in the physical world. Anything in nature that appears random and irregular may be a fractal. In fact, it could be argued that everything that exists is a fractal since it will have some structure at every level, at least down to that of an atom. Clouds, the veins in our hand, the branching of our tracheal tubes, the leaves of a tree—all show a fractal structure. In cosmology the distribution of matter across the universe is like a fractal, and its structure may descend below the atomic and nuclear level down as far as the shortest length to which any physical meaning has been ascribed, the so-called Planck length, a mere 1.6×10^{-35} meter, or about 100 million trillionth the width of a proton.

Fractals crop up not just in spatial patterns but also in temporal ones. Drumming is a case in point. It's easy to program a

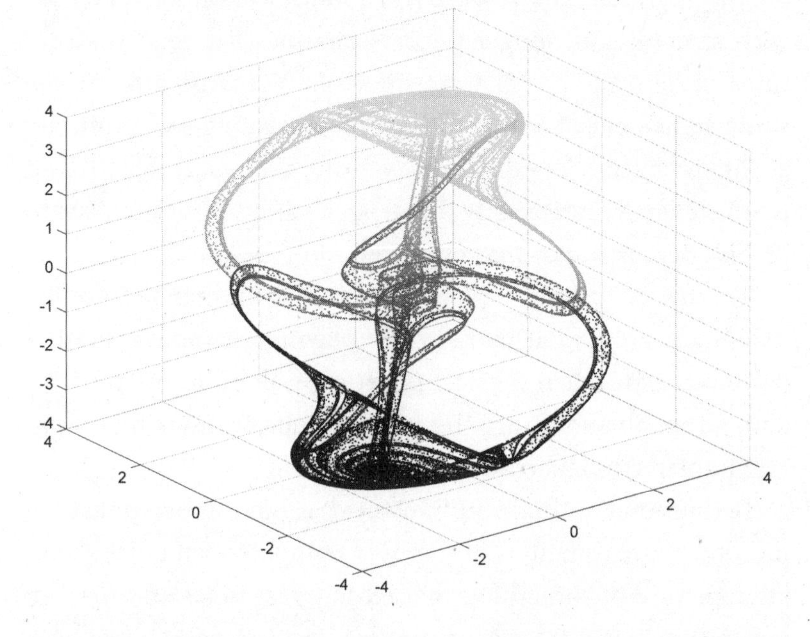

A strange attractor known as Thomas's cyclically symmetric attractor. (ANDERS SANDBERG)

computer to generate a rhythmic drum pattern or have a robot musician play one. But there's something about the sounds produced by professional drummers that distinguishes them from the perfectly steady, impeccably accurate beats of their synthetic counterparts. That "something" is the slight variations in timing and loudness—the little deviations from perfection—that, research has shown, are fractal in nature.

An international team of scientists analyzed the drumming of Jeff Pocaro, who played with the band Toto and was famed for his rapid and intricate one-handed playing of the high-hat cymbals. In both the rhythm and the loudness of Pocaro's hits on the high-hat, the researchers found self-similar patterns, with structures in longer periods of time that echoed structures present in shorter time intervals. Porcaro's hits are the sonic equivalent of a fractal coastline, revealing self-similarity at different scale lengths. What's more, the researchers found that listeners prefer exactly this type of variation, as opposed to precise percussion or that produced more randomly.

The fractal patterns differ from one drummer to another, forming part of what makes their playing distinctive. Similar patterns occur when musicians perform on other instruments and, although subtle, are the minute imperfections that separate human from machine.

Because many things in the world around us are fractals—or good approximations of them—a computer can quickly create a picture of something that closely resembles an object in nature, such as a tree. All it needs is a formula to work with and some starting data, and, in the wink of an eye, it can assemble a breathtakingly lifelike representation. Not surprisingly, then, this technique of rapidly rendering clouds, moving water, landscapes, rocks, plants, planets, and all manner of other scenery items has become a favorite of those working with CGI (computer-generated imagery) to enhance movies, animated films, flight simulators, and computer games. There's no need for vast databases to hold all the objects and scenes needed to produce a realistic moving scene when the computer can calculate it all on the fly by just cycling at high speed through a few simple rules. This approach promises to play a major role in future virtual reality and other immersive technologies where the goal is to generate 3-D imagery, indistinguishable from the actual thing in real time.

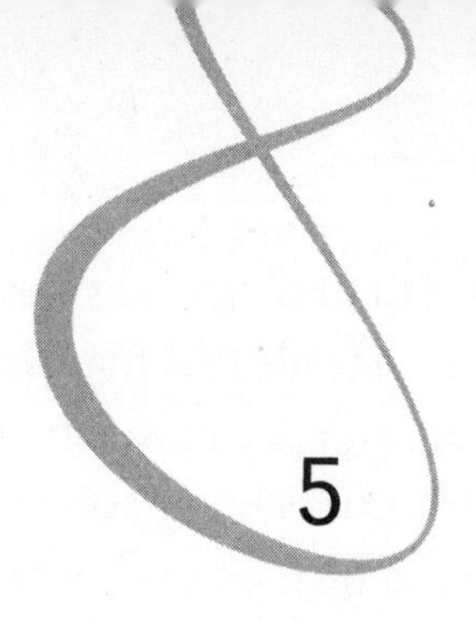

5

TURING'S FANTASTIC MACHINE

> It is possible to invent a single machine which can be used to compute any computable sequence.
>
> —ALAN TURING

Computers seem to have more in common with engineering than they do with math, and when it comes to hardware or programming applications, that's certainly true. But the theory of computation—theoretical computer science—belongs very much within the realm of mathematics. Our trek through the weird math of computers, to explore the outer limits of what it's possible to compute, begins almost a century ago, well before the first electronic brains flickered into life.

In 1928 German mathematician David Hilbert, renowned for challenging his peers with unsolved questions, posed what he called the *Entscheidungsproblem*, or "decision problem." This asked whether it's always possible to find a step-by-step procedure to decide, in a finite time, if a given mathematical statement is true. Hilbert thought the answer would turn out to be yes, but in less than a decade that hope had been dashed.

The first blow came in a paper by Austrian-born logician Kurt Gödel, published in 1931. Gödel's work, which we'll come across in more detail in the final chapter, was concerned with axiomatic systems—collections of rules, or axioms, taken to be self-evidently true—that can be used to derive theorems. Gödel showed that in any axiomatic system that's logically consistent, and big enough to encompass all the rules of arithmetic, some things are bound to be true that can't be proven to be true from within the system. What became known as Gödel's incompleteness theorems meant that there would always be some mathematical truths that were unprovable. This revelation came as a nasty shock to many but still left the door open on the question of the *decidability* of mathematical propositions—in other words, of finding a series of steps, or an algorithm, that was guaranteed to decide whether any given proposition was provable—and, if provable, whether it was true or false. Soon, though, that door was to be slammed shut as well, in part due to a young Englishman, Alan Turing, who helped give the final verdict on the *Entscheidungsproblem*.

Turing's life was a blend of triumph and tragedy—triumph because he was a genius who helped found computer science and shorten World War II, tragedy because of the way society treated homosexuals at the time. From an early age, it was clear that Turing had remarkable talents in math and science. These were in evidence at Sherborne School, in Dorset, which he began attending in 1926 at the age of thirteen. A fellow pupil was Christopher Morcom, another outstanding student, with whom Turing forged a deep friendship. The sudden death of

Morcom in 1930 had a profound effect on Turing. He threw himself even more single-mindedly into his mathematical studies and acquired, because of the loss, a keen interest in the nature of the mind and the possible survival of the spirit after death, a subject he thought might find a solution in quantum mechanics.

As an undergraduate at Cambridge, Turing took a course in logic, during which he learned about the *Entscheidungsproblem*. He decided to focus on it as part of his graduate research, having become convinced that Hilbert was wrong. There did not always exist, he believed, an algorithm for deciding whether a specific mathematical assertion could be proved. To tackle the decision problem, Turing needed a way of implementing algorithms in general, an idealized device that could carry out any logical set of instructions that was given to it. What he came up with was a purely abstract machine—he called it an *a*-machine (*a* for "automatic"), though it soon became known as a Turing machine—which he never intended should actually be built. Its design is purposely very basic, and it would be painfully slow to operate. It's meant to be just a mathematical model of a computing machine and could hardly be simpler.

A Turing machine consists of an indefinitely long tape divided into squares, on each of which may be a 1, a 0, or a blank and a read/write head. The head scans one square at a time and performs an action based on the head's internal state, the contents of the square, and the current instruction in its logbook or program. The instruction might be, for instance, "If you're in state 18 and the square you're looking at contains a '0', then change it to a '1', move the tape one square to the left, and switch into state 25."

On the machine's tape to start with is its input, in the form of a finite pattern of 1s and 0s. The read/write head is positioned over the first square of the input—say, the leftmost—and follows the first instruction that it's been given. Gradually, it works its way through the instruction list, or program, transforming the initial string of 1s and 0s on the tape into a different string until, eventually, it comes to a halt. When the machine reaches this final state, what's left on the tape is the output.

A very simple example would be adding one more to a row of n 1s; in other words, turning n into $n + 1$. The input would be the string of 1s followed by a blank square or, in the special case where $n = 0$, just a blank square. The first instruction would tell the read/write head to start at the first nonblank square, or at any square if it were known that the tape is completely blank, and read what is on the square there. If it is a 1, the instruction would be to leave it unchanged and move one square to the right while remaining in the same state; if it is a blank, the instruction would be to write a 1 in that square and stop. Having added a 1 to the string, the head might be told to stop where it is or return to the start, possibly to repeat the whole process again and add one more to the total. Alternatively, a different state could be introduced when the read/write head is positioned at the final 1 and a new program of actions continued from there.

Some Turing machines might never stop or else never stop for a given input. For example, a Turing machine that's instructed always to move to the right, whatever is in the squares it reads, will never stop, and it's easy to see this in advance.

Turing then envisioned a specific kind of Turing machine, now known as a universal Turing machine. In theory, this could run any possible program. The tape would consist of two distinct parts. One part would encode the program, while the other held the input. The read/write head of a universal Turing machine would move between these parts and carry out the program's instructions on the input. It's such a simple device: an infinitely long tape that holds both the program to be run and the input/output and a read/write head. It can carry out just six basic operations: read, write, move left, move right, change state, and halt. Yet despite this simplicity, the universal Turing machine is astonishingly capable.

You probably own at least one computer. It could have any operating system—some version of Windows perhaps, or of Mac or Android, or some other system, such as Linux. Manufacturers like to point out the advantages and special features of their own operating systems that distinguish them from the competition. However, from a mathematical standpoint, given enough memory and time, all the different operating systems are identical. What's more, they're all equivalent to a universal Turing machine, which at first glance looks far too simple to be as powerful, in terms of capability though not efficiency, as any computer in existence.

The universal Turing machine led to a concept known as emulation. One computer can emulate another if it can run a program (known as an emulator!) that effectively turns it into that computer. For example, a computer running Mac can execute a program that makes it behave as if it is running Windows—although this takes a lot of memory and is slow to process. If such an emulation is possible, the two computers are mathematically equivalent.

A programmer can also quite easily write a program to enable any computer to emulate any specific Turing machine—including a universal Turing machine (again, assuming an unlimited amount of memory is available). By the same token, a universal Turing machine can emulate any other computer by running a suitable emulator. The bottom line is that all computers, granted enough memory, can run the same programs, although it may be necessary to code them in certain specific languages, depending on how the system is set up.

There have even been various physical implementations that follow Turing's original design. These have been done mainly either as an engineering exercise or to explain how simple computation works. A number have been built using Lego, including one from a single Lego Mindstorms NXT set. In contrast, the working model created by Wisconsin inventor Mike Davey "embodies the classic look and feel of the machine presented in Turing's paper" and is on long-term display at the Computer History Museum in Mountain View, California.

Turing's actual purpose in coming up with his cunning device, as mentioned earlier, was to solve Hilbert's decision problem, which he did in 1936 in a paper titled "On Computable Numbers with an Application to the *Entscheidungsproblem*." A universal Turing machine may or may not stop given any input. Turing asked: Is it possible to determine whether it stops? You might try running it indefinitely and see what happens. However, if it went on for a long time and you chose to give up at a certain point, you would never know whether the Turing machine was going to stop right after that point or later on or whether it would carry on forever. Of course, it's possible to evaluate the outcome on a case-by-case basis, just as we can work out whether a simple Turing machine ever halts. But what Turing wanted to know was if there was a general algorithm that

A Turing machine of the form originally conceived by Alan Turing, constructed by Mike Davey. (ROCKY ACOSTA)

could decide the outcome—whether the machine stops—for all inputs. This is known as the halting problem, and Turing proved that no such algorithm exists. He then went on to show, in the final part of his paper, that this implies that the *Entscheidungsproblem* can't be solved. We can be certain that no matter how ingenious a program is, it can never work out, in all cases, whether any other program will terminate.

A month before Turing's landmark paper first appeared in print, American logician Alonzo Church, who was Turing's PhD supervisor, independently published a paper that reached the same conclusion but using a completely different approach, called lambda calculus. Like a Turing machine, lambda calculus provides a universal model of computation but more from a programming language than a hardware point of view; it deals with "combinators," which are essentially functions that act on other functions. Both Church and Turing, using their different methods, arrived at essentially the same result, which became known as the Church-Turing thesis. The gist of this is that something can be calculated or evaluated by human beings (ignoring the little matter of resource limitations) only if it's computable by a Turing machine or a device that's equivalent to a Turing machine. For something to be computable, it means that a Turing machine, given the program as input (encoded into binary), can run before ultimately terminating with the answer (similarly encoded) as output. A key implication of the Church-Turing thesis is that a general solution to the *Entscheidungsproblem* is impossible.

Although Turing invented his machine to solve a mathematical problem, he effectively laid down a blueprint for the development of digital computers. All modern computers basically do what Turing machines do, and the concept is also used to measure the strength of computer instruction sets and programming languages. These are said to be Turing-complete, and therefore at the pinnacle of computer program strengths, if they can be used to simulate any single-tape Turing machine.

No one has yet come up with a way of computing things that can do more than what a Turing machine does. Recent developments in quantum computers seem, at first sight, to offer a way of transcending the Turing machine's powers. But in fact, even a quantum computer, given enough time, can be emulated by any ordinary (classic) computer. For some types of problems, quantum computers may be much more efficient than any classic equivalent, but, in the end, everything of which they're capable can be done on the simple device that Turing envisioned. What this shows is that there are some things we can't hope to compute and get a general answer to that's guaranteed to be accurate (although we may be able to do it on a case-by-case basis).

There are other things in math that, superficially, look nothing like Turing machines but turn out to be equivalent to them by emulation. One example is the Game of Life, devised by English mathematician John Conway. The game came about because of Conway's interest in a problem investigated in the 1940s by American mathematician and computer pioneer John von Neumann. Could a hypothetical machine be contrived that could make exact copies of itself? Von Neumann found that it could by creating a mathematical model for such a machine using very complicated rules on a rectangular grid. Conway wondered if there was a much simpler way of proving the same result and so arrived at the Game of Life. Conway's game is played on an (in theory) infinite square grid of cells that can be colored either black or white. Some starting pattern of black cells is laid down and then allowed to evolve according to two rules:

1. A black cell remains black if exactly two or three of its eight neighbors are black.

2. A white cell turns black if exactly three of its neighbors are black.

That's all there is to it. Yet despite the fact that a child could play it, Life has all the capability of a universal Turing machine—and therefore of any computer that has ever been built. Conway's remarkable game was first brought to the attention of the wider world through Martin Gardner's "Mathematical Games" column in the October 1970 edition of *Scientific American.* Gardner introduced his reader to some of the basic patterns in Life, such as the "block," a single two-by-two black rectangle, which under the rules of the game never changes, and the "blinker," a one-by-three black rectangle, which alternates between two states, one horizontal and the other vertical, keeping a fixed center. The "glider" is a five-unit shape that moves diagonally by a distance of one square every four turns.

Conway originally thought that no pattern that was chosen at the outset would grow indefinitely—that all patterns would eventually reach some stable or oscillating state or die out altogether. In Gardner's 1970 article on the game, Conway issued a challenge with a fifty-dollar reward for the first person who could either prove or disprove this conjecture. Within weeks the prize had been claimed by a team from the Massachusetts Institute of Technology led by mathematician and programmer Bill Gosper, one of the founders of the hacker community. The so-called Gosper glider gun, as part of its endless, repetitive activity, spits out a steady stream of gliders at a rate of one per thirty generations. As well as being fascinating to watch, this also turns out to be of great interest from a theoretical standpoint. Ultimately, the Gosper guns are crucial for building computers in Life because the streams of gliders they emit can be considered analogous to electrons in a computer. In a real computer, though, we need ways of controlling these streams in order to actually compute, which is where logic gates come in.

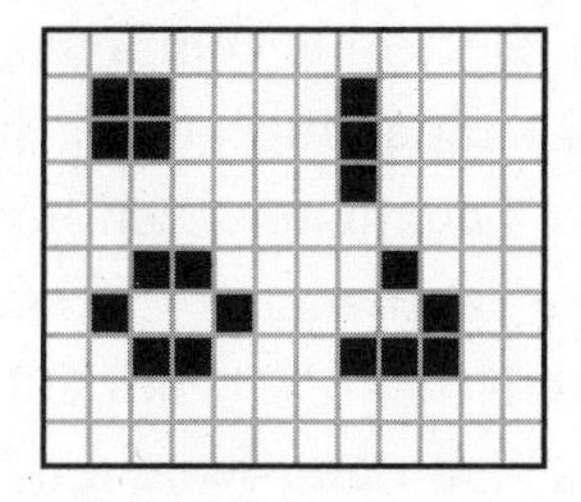

Four common Game of Life patterns. On the left are a "block" *(top)* and a "beehive" *(bottom)*. These are both "still-lifes," which means that they stay the same in each generation. The top-right pattern is a "blinker," the most common oscillator, which returns to its starting position after a number of generations. In this case, it alternates between a vertical and a horizontal form. The bottom-right pattern is a "glider." (AGNIJO BANERJEE)

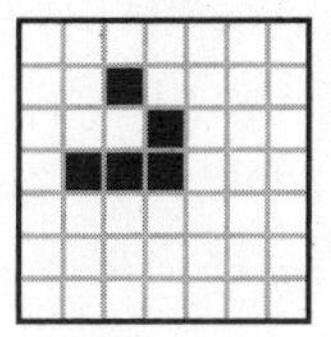

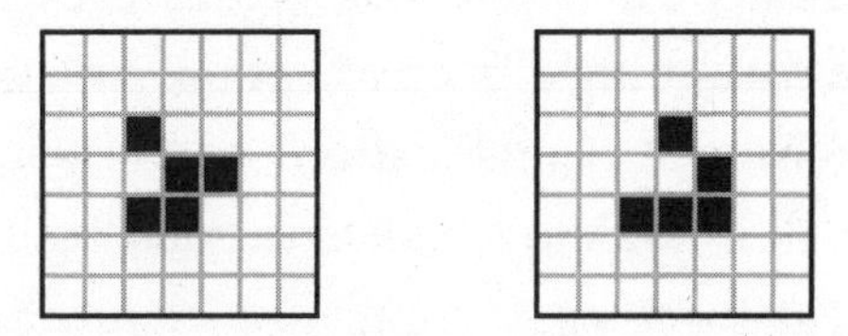

A glider, in four generations, moves diagonally by one cell. (AGNIJO BANERJEE)

A logic gate is an electronic component that takes in one or more signals as an input and gives a signal as an output. It's possible to build a computer from just one type of logic gate, but using three types makes the task a lot easier. The gates in question are NOT, AND, and OR. A NOT gate outputs a high signal as an output if and only if it receives a low signal as an input. An AND gate outputs a high signal as an output if and only if both its inputs are high. An OR gate outputs a high signal if and only if at least one input is high. These gates can be combined to form circuits that can both process and store data.

An infinite circuit of logic gates can be used to simulate a Turing machine. In turn, logic gates can be simulated by patterns in Conway's Game of Life—specifically, by using various combinations of Gosper glider guns. A stream of gliders issuing from one of these guns can represent a "high" signal (a 1) and an absence of gliders a "low" signal (a 0). Crucially, one glider can block another one because if two gliders meet in the right way, they annihilate. The final piece of the puzzle is something known as an "eater," which is a simple configuration of seven black cells. An eater can absorb excess gliders, thereby preventing them from disrupting other parts of the pattern, while itself remaining unchanged. Combinations of Gosper glider guns and eaters are all it takes to simulate various logic gates, which can then be put together to simulate a complete Turing machine. So, remarkably, there's nothing that even the world's most powerful supercomputer can do that, given enough time, can't be computed using the Game of Life. It's also impossible to write a program that will predict the fate of any arbitrary Life pattern, as such a program would then be able to solve the halting problem. The Game of Life, like life itself, is unpredictable and full of surprises.

The modern field of computation theory is built on Turing's ideas but now also embraces another concept that Turing did not consider. In his famous 1936 paper, Turing was concerned only with the existence of algorithms, not their efficiency. But in practice, we also want our algorithms to be fast so that computers can solve problems as quickly as possible. Two algorithms might be equivalent—that is to say, they can both solve the same problem—but if one takes a second and the other a million years, we would obviously choose the former. The trouble with quantifying the speed of algorithms is that it depends on many factors, to do with both hardware and software. For example, different programming languages may result in different execution speeds for the same set of instructions. Computer scientists commonly use Big O (*O* for "order") notation to quantify speed in relation to the size of the input (n). If a program runs in order n, or $O(n)$ time, this means that the time taken for the program to run is roughly proportional to the size of the input. This is the case, for example, when adding two numbers together using decimal notation. Multiplying numbers, however, takes longer—$O(n^2)$ time.

If a program runs in something called polynomial time, it means that the time taken is no more than the input size raised to a fixed power. This is generally considered fast enough for most purposes. Of course, if the power is huge—say, the one hundredth power—the program would take far too long to run, but this hardly ever happens. One example of an algorithm with a reasonably high power is the Agrawal-Kayal-Saxena algorithm, which can be used for testing whether a number is prime. This runs in $O(n^{12})$ time, so in most cases a different algorithm is used, which runs slower than polynomial time but is faster than the AKS algorithm for many practical values. In searching for new, very large primes, however, the AKS algorithm comes into its own.

Suppose we take a very simple-minded approach to determining whether a number, consisting of *n* digits, is prime. This would involve checking all the numbers from 2 up to the square root of the given number to see if they were factors. We can exploit a few shortcuts, such as skipping even numbers, but still the time taken to test for primality this way comes out to be $O(\sqrt{10^n})$, or roughly $O(3^n)$. This is exponential time, which is manageable using a computer providing that *n* is fairly small. Testing a 1-digit number for primality using this method involves 3 steps, which, assuming 1 quadrillion steps per second (a typical speed for a supercomputer), would take 3 femtoseconds (3 quadrillionths of a second). A 10-digit number would take about 60 picoseconds to check and a 20-digit number about 3.5 microseconds. But running on exponential time, a program eventually gets hopelessly bogged down. A 70-digit number, using our primitive method, would take about 2.5 quintillion seconds to check, which is far longer than the present age of the universe. Fast algorithms prove their worth in such situations.

Using something like AKS, assuming the time taken is the twelfth power of the input size, a 70-digit number takes "only" 14 million seconds, or 160 days, to check for primality. This is still a long time for a high-speed computer run, but is like the wink of an eye compared with the cosmic timescales demanded by an exponential algorithm. Polynomial algorithms may or may not be practical in reality, but exponential algorithms are certainly not practical when dealing with large inputs. Fortunately, there's also a wide range of algorithms between the two, and it's often the case that algorithms that are nearly polynomial work well enough in practice.

The Turing machines we've talked about so far have all had one important thing in common. The lists of rules—the algorithms—telling them what to do have always prescribed just one action to be carried out in any situation. Such Turing machines are said to be deterministic Turing machines (DTMs). When given an instruction, they mechanically follow that instruction; they cannot "choose" between two different instructions. However, it's also possible to conceive of another type, known as nondeterministic Turing machines (NTMs), which, for any given state of the read/write head and any given input, allow more than one instruction to be carried out. NTMs are just thought experiments—it would be impossible to actually build one. For example, in its program, an NTM might have both "If you are in state 19 and see a '1,' change it to a '0' and move one place to the right" and "If you are in state 19 and see a '1,' leave it unchanged and move one place to the left." In this case, the internal state of the machine and the symbol being read on the tape don't specify a unique action. The question, then, is how does the machine know which action to take?

An NTM explores all the possibilities for solving a problem and then, at the end, chooses which, if any, is the correct answer. One way to think of this is that the machine is an exceptionally lucky guesser that always manages to pick the right solution. Another, and perhaps more reasonable, way to picture an NTM is as a device that gains in computational power as it goes along, so that whatever is thrown at it in each step of the calculation takes no longer to process than the previous step. The task in hand might be, for instance, to search a binary tree—an arrangement of data that at each point or node splits off into two more options. Suppose the object is to find a certain number—say, 358—in the tree. The machine has to go down every possible route until it comes across this value. An ordinary Turing machine—a DTM—would have to go down every possible path through the tree, one after another, until it hit on the target value. Because the number of branches increases exponentially, doubling at every level of the tree, the time taken to locate the node containing 358 would be hopelessly long unless, by good fortune, it didn't lie too far down the tree. With an NTM available, however, the situation changes dramatically. As each level of the binary tree is reached, the NTM can be imagined to double in processing speed, so that it searches every level of the tree in the same amount of time, no matter how many nodes there are.

NTMs are devices of the imagination—thought experiments that can be used to explore interesting problems in computation. They can't be built in practice, although, in principle, everything an NTM can do a DTM could also do, given enough time. But that "enough time" is the catch. An NTM could do in polynomial time what would take a DTM exponential time. Too bad we can't ever actually build one. What these computers of the imagination let us do, however, is come to grips with one of the great unsolved problems of computer science and of mathematics as a whole: the so-called P versus NP problem. One million dollars, from the Clay Mathematics Institute, awaits the first researcher to come up with a provably correct solution. P and NP are the names given to two sets of problems having different classes of complexity. Problems in set P (polynomial) are those that can be solved by an algorithm running in polynomial time on an ordinary (deterministic) Turing machine. Problems in set NP (nondeterministic polynomial) are those that we know how to solve in polynomial time if we had access to an NTM. (Factoring large numbers is one such problem. An NTM can search through the binary tree for the "right" factor rapidly, in polynomial time, whereas a DTM has to search every single branch, taking exponential time.) What this means is that every problem in P is also in NP, as an NTM can do anything an ordinary Turing machine can do in the same amount of time.

It seems reasonable to suppose that NP is bigger than P because it includes problems that are known to be tractable only on a Turing machine with superpowers: one that's enhanced by making it outrageously lucky or ridiculously fast. However, there is, as yet, no proof that a regular DTM cannot do everything an NTM can, even though it seems highly likely. To mathematicians, though, there's a world of difference between reasonable supposition and certainty. Until it's demonstrated otherwise, it remains possible that someone will prove that the sets P and NP are equal, which is why it's called the P versus NP problem. One million dollars is a handsome prize, but how could anyone ever claim it when it means proving (or disproving) that *all* NP problems are P? A small ray of hope is offered by certain problems in NP that are said to be NP-complete. NP-complete problems are remarkable in that if a polynomial algorithm could be found that would run on an ordinary Turing machine and solve any one of these problems, then it would follow that there's a polynomial algorithm for every single problem in NP. In this case, P = NP would be true.

The first NP-complete problem was found by American Canadian computer scientist and mathematician Stephen Cook in 1971. Known as the Boolean satisfiability problem, or SAT, it can be expressed in terms of logic gates. It starts with an arrangement of arbitrarily many logic gates and inputs (but no feedback) and exactly one output. It then asks whether there's some combination of inputs that will turn the output on. A solution could always be found, in principle, by testing every possible combination of inputs to the entire system, but this would be equivalent to an exponential algorithm. To show that P = NP, it would have to be shown that there was a faster—polynomial—way of getting the answer.

While SAT was the first NP-complete problem to be identified, it isn't the most famous. That honor belongs to the Travelling Salesman problem, which had its origins in the middle part of the nineteenth century. A manual for traveling salesmen published in 1832 talked about the most effective way to tour locations in Germany and Switzerland. The first academic treatment came a decade or two later from Irish physicist and mathematician William Hamilton and Church of England minister and mathematician Thomas Kirkman. Suppose a salesman has to travel to a large number of cities and knows the distance, which doesn't necessarily have to be straight, between each pair of cities. The problem is to find the shortest path that visits all of the cities and returns to the start. Only in 1972 was it shown that the Travelling Salesman problem is NP-complete (meaning that a polynomial algorithm for this problem would prove P = NP), which explained why generations of mathematicians, latterly even using computers, had trouble finding optimal solutions for complicated routes.

The Travelling Salesman problem may be easy to grasp, but it's no easier to solve than any other NP-complete problem: they're all as hard as each other. Mathematicians are tantalized by the fact that finding a polynomial algorithm for any NP-complete problem would prove that P = NP. Such a proof would have serious implications, including that there would be a polynomial algorithm for cracking RSA, the method of encryption, described later, that we rely upon on a daily basis—for example, for banking. But in all likelihood, none exists.

While nondeterministic Turing machines exist only in the mind, quantum computers, which are also potentially very powerful, are in the early stages of development. As their name suggests, they make use of some of the very strange goings-on in the realm of quantum mechanics and work not with ordinary bits (binary digits) but with quantum bits, or "qubits." A qubit, which can be as simple as an electron with an unknown spin, has two properties due to quantum effects that an ordinary bit in a computer does not. First, it may be in a superposition of states. This means a qubit may represent both a 1 and a 0 at the same time and resolves into one or the other only when it's observed. Another way of interpreting this is that the quantum computer, along with the rest of the universe, splits into two copies of itself, one of which has the bit 1 and the other the bit 0. Only when the qubit is measured does it, and the universe around it, coalesce to a specific value. The other curious property upon which quantum computers depend is entanglement. Two entangled qubits, though separated in space, are linked by what has been called "spooky action-at-a-distance"

Professor Winfried Hensinger (*left*) and Dr. Seb Weidt (*right*) working on a quantum computer prototype. (ION QUANTUM TECHNOLOGY GROUP, UNIVERSITY OF SUSSEX)

so that measuring one instantaneously affects the measurement for the other.

Quantum computers are computationally equivalent to Turing machines. But as we've seen, there's a difference between being able to compute something at all (given enough time) and being able to compute it efficiently. Anything a quantum computer can do (or will be able to do) is also achievable on a classical paper-tape Turing machine if we're prepared to wait a few geological eras or more. Efficiency is a different matter altogether. For some types of problem, quantum computers are likely to be many times faster than today's conventional computers, although, in terms of what is actually computable, all have exactly the same capability as Turing's original design.

It's tempting to think that quantum computers are the same as nondeterministic Turing machines, but this isn't the case. Computationally, they're equivalent, because nondeterministic Turing machines can't surpass deterministic Turing machines in terms of what is computationally possible (you could program a DTM to simulate either of them). In terms of efficiency, however, quantum computers are suspected to be inferior to NTMs—not surprisingly, since NTMs are entirely devices of the imagination. In particular, it's unlikely, although it remains to be seen, whether they'll be able to solve NP-complete problems in polynomial time. One problem that *has* been solved in polynomial time with quantum computers, which was previously thought to have no such solution (assuming P = NP is false), is the factoring of large numbers. In 1994 American applied mathematician Peter Shor found a quantum algorithm that makes use of the special properties of the problem. Unfortunately, similar techniques can't be applied to other problems, such as those known to be NP-complete. If there's any polynomial algorithm to solve an NP-complete problem with quantum computing, it will again have to exploit specific features of the situation.

Quantum computers, like most new upstart technologies, bring both hope and headaches. Among the latter is the possibility of cracking codes that were previously thought to be highly secure, mainly because, despite decades of research, no one has found any polynomial-time method of cracking them. Modern encryption methods are based on an algorithm known as RSA, which is an acronym of its inventors, Ron

Rivest, Adi Shamir, and Leonard Adleman. Using the algorithm to encrypt data can be done very quickly and happens many times, every second of every day, during online data transactions. However, applying RSA in reverse, to decrypt data, is extremely slow, requiring exponential time, unless special information is provided. This asymmetry of speed and the need for special information explain why RSA is so effective. The way RSA works is that each person using the system has two keys, a public key and a private key. The public key allows encryption and can be known to everyone, whereas the private key enables decryption and is known only to the owner of the keys. Sending a message is easy because it involves just applying the algorithm with the public key. However, the message can be read only by the intended recipient, who knows the private key. It's theoretically possible to work out the private key if given the public key, but it depends on being able to factor huge numbers with hundreds of digits. If the keys are large enough, it would take all the computers in the world, working together far longer than the current age of the universe, to decrypt the messages we send daily during our banking and other confidential transactions. Quantum computers, however, threaten to change all this.

In 2001 Shor's algorithm, a method of factoring numbers in polynomial time, was used with a seven-qubit computer to factor 15 into 3 × 5. A decade later, the factorization of 21 was achieved by the same method. Both achievements may seem comically underwhelming, given that any child who knows their times tables could do the same without much delay. However, in 2014, a different technique in quantum computing was used to find the prime factors of much larger numbers, the biggest of which was 56,153. Even this may not seem very impressive when faced with trying to factorize numbers with hundreds of digits. But as quantum computers with more and more qubits become available, it's only a matter of time before it becomes possible to crack, efficiently, all RSA ciphers. When this happens, the current method of handling online transactions will no longer be secure, and the banking industry, along with every other aspect of modern life that depends on the secure transfer of data, will be thrown into chaos. Perhaps it will be possible to develop a new encryption system based on an NP-hard problem: in other words, a problem at least as hard as an NP-complete problem, though not necessarily belonging to the NP set. NP-complete problems are very difficult to solve in the worst-case scenarios, but good algorithms can usually be found in more typical cases. They would give an encryption method that is generally easy to crack, but with a small probability of being extremely difficult. What's needed is something that is almost always extremely difficult, taking exponential time or longer, to crack. No encryption method like this has yet been discovered, though it remains a possibility. If quantum computers can't crack NP-complete (and therefore NP-hard) problems, then if one were ever discovered, we might have security again—at least for a while.

Most computer scientists suspect that P ≠ NP. It's a belief bolstered by decades of research that have failed to find a single polynomial-time algorithm for solving any of the more than three thousand significant known NP-complete problems. The failure-to-date argument, however, isn't very convincing, especially in light of the unexpected proof of Fermat's last theorem—a simply stated problem that took intensive effort and cutting-edge methods to resolve. Nor is it particularly convincing to believe that P ≠ NP purely on philosophical grounds. Theoretical computer scientist Scott Aaronson of MIT has said, "If P = NP, then the world would be a profoundly different place than we usually assume it to be. There would be no special value in 'creative leaps,' no fundamental gap between solving a problem and recognizing the solution once it's found." However, both math and science have shown themselves to be perfectly capable of blindsiding us and transforming our intellectual worldview almost overnight. If it did turn out that P = NP, then, to begin with, there might be little practical impact, because a proof, if it existed, would most likely be nonconstructive. In other words, although a proof might show that polynomial algorithms existed for NP-complete problems, it wouldn't be able to actually describe any. For a while at least, our encrypted data would remain secure—though for how long would be uncertain once major mathematical efforts began in search of such an algorithm.

In any event, before the security of our data is threatened by any developments on the P versus NP problem or more efficient algorithms, quantum mechanics may come to our rescue. The field of quantum cryptography may result in a cipher that is completely unbreakable, no matter what decryption techniques are brought to bear on it. An example of a truly unbreakable cipher was found as long ago as 1886 and is known as the onetime pad. The key is a random sequence of letters that is as long as the message. The message is combined with the key by converting letters into numbers ($A = 1$, $B = 2$, and so on), adding the corresponding numbers for each letter of the message and key, subtracting 26 if the sum is greater than 26, and converting back to letters. This has been proven to be completely unbreakable. Even if someone had enough time to try out every combination, they would be completely unable to tell the correct message apart from every possible incorrect message. The whole setup, however, depends on the key being destroyed after use, because if it were reused, then anyone who got hold of both encrypted messages would be able to decrypt them if they knew that the key had been reused. The keys must also be exchanged privately, as anyone who gained access to the key could instantly decrypt the supposedly secure message. Onetime pads were once used by Soviet spies and kept in tiny books that were highly flammable to aid in their secure destruction and are still used in the hotline between the presidents of the United States and Russia. But the need to exchange keys privately and securely is a big disadvantage and makes the method impractical for most purposes, such as online transactions.

Quantum mechanics promises to change all this. It relies on the fact that measuring a certain property of light particles, or photons, known as polarization, affects the polarization. (Polarization describes the way in which the waves associated with photons vibrate at right angles to the direction of their travel.) The crucial fact is that if the polarization is measured twice in the same direction, the result will be the same. One method of doing the measurement uses a type of filter called an orthogonal filter. If the light is polarized vertically or horizontally, it will pass through an orthogonal filter and retain its polarization. If it starts off polarized any other way, the light will still pass through, but its polarization will change to become either vertical or horizontal. Another method of doing the polarization measurement is using a diagonal filter, which works in a similar way but with light vibrating midway between horizontal and vertical. The final components of the cryptography system are two more filters. One of these tests whether light that has passed through the orthogonal filter is horizontally or vertically polarized; the other performs the same test on light that's gone through the diagonal filter, testing in which diagonal direction the light is polarized.

Suppose we wanted to send a random bit for use in a onetime pad. We would send a photon through either the orthogonal or the diagonal filter, chosen randomly, and then record whether it was polarized vertically or horizontally. We would ask the recipient to do the same. They would then tell us which filter they had used, and we would confirm which filter we had used. If they were the same, this bit could be stored for later use in a onetime pad. If not, the bit would be thrown out and the process repeated. An eavesdropper wouldn't be able to tell which filter was used until after that photon had passed through the system and they were no longer able to measure it. Moreover, since measuring the polarization could change it, after we had enough bits we could compare a small number of them, discarding them afterward. If they all matched, our channel could be assumed secure and the rest of the bits safely used in a onetime pad. If not, it would demonstrate the presence of an eavesdropper, and all of the bits would be rejected as useless. So quantum cryptography not only protects a onetime pad from eavesdroppers but can also detect that an attempt at eavesdropping took place—something beyond the capability of conventional cryptography.

Progress in quantum computing is now very rapid. In 2017 physicists at the University of Sussex published construction plans for future large-scale quantum computers, thereby making the design freely available to all. The Sussex blueprint shows how to avoid a problem, known as decoherence, that had blighted previous laboratory attempts to build devices with more than ten or fifteen qubits. It also describes some specific technologies that would help to make powerful quantum computers, with much larger numbers of qubits, a reality. These include the use of room-temperature ions (charged atoms), confined in traps, to serve as qubits; the application of electric fields to push the ions from one module of the system to another; and logic gates that are controlled by microwaves and variations in voltage. The Sussex team next intends to build a small prototype quantum computer. Meanwhile, other groups at Google, Microsoft, and various start-ups, such as IonQ, are pursuing their own schemes, based on the trapped-ion approach, superconductivity, or (in Microsoft's case) a design known as topological quantum computing. IBM has announced its intention to market a fifty-qubit quantum computer "in the next few years," and scientists are already looking ahead to the time when machines with millions or billions of qubits become a reality.

Had Turing been alive today, he would no doubt have been involved with the latest developments in computation, including, very possibly, theoretical work on quantum computers. He would have avoided the primitive attitudes toward sexuality that were pervasive during his life and surely contributed to his early death. But one thing that he would have found unchanged is the concepts of algorithms and universal computation that, through his remarkable, and remarkably simple, machine, he played such a major part in developing.

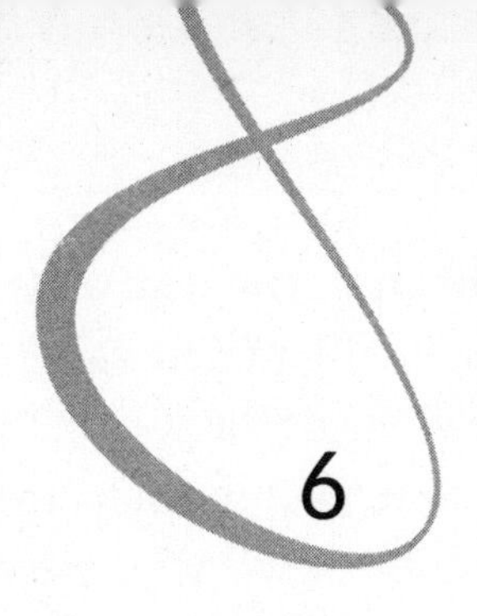

6

MUSIC OF THE SPHERES

> May not music be described as the mathematics of the sense, mathematics as music of the reason? The musician feels mathematics, the mathematician thinks music: music the dream, mathematics the working life.
>
> —JAMES JOSEPH SYLVESTER

Music, at its very core, is mathematical. It's often said that mathematics is a universal language that could be used as a first means of communication between intelligent species on different worlds. But that claim of universality might also apply to music, and, in fact, we've already sent some of our music toward the stars in the hope that beings out there might hear it and so come to understand something of the creatures who made it.

Voyager 1, launched on September 5, 1977, recently became the first human-made object to enter interstellar space. Having flown past Jupiter and Saturn, it headed out of the solar system and in 2012 passed beyond the heliopause, the boundary where the influence of the sun's magnetic field ends and that of the rest of the galaxy begins. Its sister craft, *Voyager 2*, launched in the same year, is also heading for the void between stars but in a different direction. Both remain in contact with earth, sending back data from a handful of science experiments that their dwindling power reserves can supply, but neither is destined for any close encounters with another star system in the foreseeable future. Their speed is so small compared with the immensity of interstellar distances that it would take them tens of thousands of years to reach even the nearest star, Proxima Centauri—assuming they were heading directly toward it (which they aren't).

According to NASA's current estimates, *Voyager 1* will come within 1.6 light-years of the star Gliese 445 and *Voyager 2* within 1.7 light-years of Ross 248 about 40,000 years from now. By the time of these very remote flybys, both probes will be long dead. But structurally, the *Voyager*s could remain intact for millions of years, drifting through the Milky Way galaxy and—who knows—possibly be found by some advanced race that would be curious about the probes' origins and creators. In that unlikely event, each spacecraft carries a message in the form of a gold-plated copper phonograph record containing sounds and images intended to portray the variety of life, environments, and human cultures on earth. As well as 116 pictures, a variety of natural sounds, and spoken greetings in fifty-seven different languages, the *Voyager* Golden Record features ninety minutes of music from different ages and regions of the world, including passages from Stravinsky's *The Rite of Spring*, a gamelan piece from Indonesia, Bach's Brandenburg Concerto no. 2, and Chuck Berry's "Johnny B.

The *Voyager* Golden Record. (NASA/JPL)

Goode." Thoughtfully, a stylus and coded instructions are provided by which to play the record. But assuming aliens ever found one of the golden records and managed to play the music as intended, the question is whether they would recognize it for what it is. And, similarly, if alien music somehow reached our ears, would we appreciate it as being musical?

One of us (David) is a singer and songwriter whose album *Songs of the Cosmos* combines science with music in tunes such as "Dark Energy." But as well as songs with a scientific flavor, there's also science in the making of music and mathematics, deeply rooted, in the relationship between notes and the construction of scales.

It was the ancient Greeks who first discovered that there's a strong link between music and mathematics. Pythagoras and his followers, in the sixth century BC, built an entire cult around their belief that "all is number" and that the whole numbers were especially significant. Each of the numbers 1 to 10, they held, had a unique significance and meaning—1 was the generator of all other numbers, 2 stood for opinion, 3 for harmony, and so on, all the way up 10, which was the most important and known as *tetraktys*, because it is the triangular number made from the sum of the first four numbers, 1, 2, 3, and 4. Even numbers were considered female and odd numbers male. In music the Pythagoreans delighted in their discovery that the most harmonious-sounding intervals corresponded with whole-number ratios. The very same numbers they held in such high esteem on an intellectual level determined, as simple fractions, what set of notes was most satisfying to the ear. A vibrating string held down at its halfway point (2:1) sounds an octave higher than when open. Held down and played so that the length of the vibrating section to the entire string is in the ratio 2:3 gives a perfect fifth (so-called because it is the fifth note in the scale and highly consonant with the root note). Likewise, 4:3 produces a perfect fourth and 5:4 a major third. Since frequency depends on one divided by the string length, these ratios also give a relationship between the frequencies of the notes.

The simplest of the ratios (apart from the octave)—the perfect fifth—is the basis for what's become known as Pythagorean tuning because modern musicologists ascribe its origins to Pythagoras and his brotherhood. Start with a note such as D and move up by a perfect fifth and down by a perfect fifth to produce other notes of the scale, A and G, respectively. Now move up another perfect fifth from A and down another perfect fifth from G to generate the next notes, and so on.

Eventually, we end up with an eleven-note scale centered on D, like this:

E♭–B♭–F–C–G–D–A–E–B–F♯–C♯–G♯

Without some adjustments, this would span a wide frequency range, equivalent to seventy-seven notes on a piano. To make the scale more compact, low notes are shifted to a higher octave, by having their frequencies doubled or quadrupled, and higher notes are similarly shifted downward by an octave or two. The result of bringing the notes closer together in this way is what's called the basic octave. Pythagorean tuning was used by musicians in the West until about the end of the fifteenth century, when its limitations for playing a wider variety of pieces became apparent.

So enamored were the Pythagoreans with their discovery—that simple ratios of vibrating strings equated to harmonious musical intervals—and their belief that the universe was based on whole numbers that they saw a perfect marriage of music and mathematics in the heavens. At the center of physical space, according to their cosmology, was a great fire. Around this, carried on transparent celestial spheres and moving in circular paths, were ten objects, in order from the center: a counter-Earth, Earth itself, the moon, the sun, the five known planets or "wandering stars" (Mercury, Venus, Mars, Jupiter, and Saturn), and, finally, the fixed stars. The separations between these spheres, they taught, corresponded to the harmonic lengths of strings, so that the movement of the spheres gave rise to a sound (inaudible to human ears) known as the "harmony of the spheres."

Both of the Greek words *harmonia* (meaning "joint" or "agreement") and *arithmos* ("number") come from the same Indo-European root, *ari*, which also crops up in English words such as "rhythm" and "rite." Harmonia was also the Greek goddess of peace and harmony—fittingly, since her parents were Aphrodite (goddess of love) and Ares (god of war). The Pythagorean idea that musical harmonies were inherent in the spacing of heavenly bodies persisted throughout the Middle Ages. The philosophy of Musica Universalis (Universal Music) found its way into the quadrivium, a quartet of academic subjects, arithmetic, geometry, music, and astronomy, that was taught after the trivium (grammar, logic, and rhetoric) in medieval European universities and was based on Plato's curriculum for higher education. At the heart of the quadrivium was the study of number in various forms: pure number (arithmetic), number in abstract space (geometry), number in time (music), and number in both space and time (astronomy). Following the lead of Pythagoras, Plato saw an intimate connection between music and astronomy: music expressing the beauty of simple numerical proportions to the ears and astronomy to the eyes. Through different senses, they expressed the same underlying unity, based on mathematics.

More than two thousand years later, German astronomer Johannes Kepler took the notion of a musical cosmos a step further, by linking together fundamental shapes and melodic sounds across the heavens. Kepler believed in astrology and was devoutly religious, as were many other intellectuals of his time, but he was also a key figure in the scientific revolution of the Renaissance. He's best remembered for his three laws of planetary motion, built on the foundation of accurate

Fig. 37 —Kepler's Analogy of the Five Solids and the Five Worlds.

Kepler thought that the orbits of the then-known planets were spaced to match a nesting of the Platonic solids. From the book *The Science-History of the Universe*, by Francis Rolt-Wheeler. (NEW YORK: *CURRENT LITERATURE*, 1910)

observations of the planets by Danish nobleman Tycho Brahe. Early in his career, Kepler was fascinated by the notion that there might be a geometric basis to the spacing of the planets. To the sun-centered model of the solar system proposed earlier by Polish astronomer Nicolaus Copernicus, Kepler, in his 1597

Mysterium Cosmographicum (*The Cosmographic Mystery*), added the idea that the five Platonic solids—the only regular, convex polyhedrons in 3-D—held the key to the spacing of worlds. By inscribing and circumscribing with spheres these solids in a certain order—octahedron, icosahedron, dodecahedron, tetrahedron, and cube—Kepler believed he could generate the orbs within which the six known planets (Mercury, Venus, Earth, Mars, Jupiter, and Saturn) moved. God, it seemed, might not be a numerologist, as the Pythagoreans believed, but a geometer.

Going beyond mere speculation, Kepler carried out acoustic experiments at a time—the dawn of the seventeenth century—when testing ideas in practice was still a novel concept in academic circles. Using a monochord, he checked the sound made by the string when stopped at different lengths and established by ear what divisions were most pleasing. As well as the fifth, which was of all-consuming importance to the Pythagoreans, he noted that the third, the fourth, the sixth, and various other intervals were also consonant. He wondered if these harmonious ratios might be reflected in the heavens, so that the old notion of the harmony of the spheres might be brought up-to-date and more in line with the latest observations. Perhaps the ratio of the greatest and least distances between planets and the sun matched some of the consonant intervals he had found. But no, they did not. He considered then the *speed* of the planets at the points of maximum and minimum distance, where he knew, from observations, that they moved the slowest and the fastest, respectively, in relation to the sun. Movement, he noted, would be a better analogue than distance to the vibration of a string, and, indeed, using this planetary property, he found what seemed to be a connection. In the case of Mars, the ratio of its extreme orbital speeds (measured in terms of angular motion across the sky) was about 2:3, equivalent to a perfect fifth, or "diapente," as it was known until the late nineteenth century. The extreme motions of Jupiter differed by a ratio of about 5:6 (a minor third in music) and those of Saturn by very close to 4:5 (a major third). The corresponding ratios for Earth and Venus were 15:16 (roughly the difference between *mi* and *fa*) and 24:25, respectively.

Encouraged by these correspondences, which, as it turned out, were fortuitous, Kepler went in search of more subtle cosmic harmonies. He looked at the ratios of the speeds of neighboring worlds and convinced himself that harmonious ratios underpinned not only the movement of planets individually but also how they moved relative to one another. All of these thoughts on the subject he wrapped into a grand unified theory of how consonant intervals in music were linked to movements in the heavens and published it in his magnum opus, *Harmonices Mundi* (*The Harmony of the World*), in 1619.

Shortly after, he made a discovery that today is known as his third law of planetary motion. He found a precise connection between the time it takes a planet to go once around the sun and its distance from the sun—namely, the square of a planet's period is proportional to the cube of its semimajor axis. This is the relationship still taught in physics classes today, but it was uncovered originally during the course of Kepler's mystical studies into the harmonic structure of the cosmos.

Kepler helped propel astronomy into the modern era with his crucial insight that the orbits of planets aren't circular, as the ancients had believed, but elliptical. This paved the way for

Newton's universal theory of gravitation, but, less obviously, it also set the stage for innovative and more flexible systems of tuning in music. From his experiments in auditory space, Kepler wondered if there was a smallest interval—a lowest common factor—from which all other harmonies could be built. He found that there wasn't. Just as planetary orbits weren't based on perfect circles, there was no neat and simple way to achieve musical consonance using one fundamental interval. This became most obvious when any attempt was made to change the key of a piece of music.

Pythagorean tuning, based on stacked fifths, is one example of what's called *just tuning*, in which the frequency of notes is related by ratios of reasonably small whole numbers. If we take the scale of C major, for example, divide it up into eight pitches (CDEFGABC), and give the tonic or root note, C, the ratio 1:1, and the fifth, G, the ratio 3:2, in Pythagorean tuning the notes above C have the following frequency ratios relative to C: D, 9:8; E, 81:64; F, 4:3; G, 3:2; A, 27:16; B, 243:128; and C (one octave up), 2:1. This arrangement works fine, providing we stay in the same key or use flexible instruments, such as the human voice, which can make fine adjustments to intonation on the fly. But any form of just tuning runs into problems with instruments like the piano, which, once tuned, can produce only certain frequencies.

Composers and musicians earlier than Kepler had started to break out of the rigid confines of Pythagorean tuning. But it was around Kepler's time that the first important moves, in Europe at least, were made away from the notion of just tuning altogether. A pioneer of the new trend was Galileo's father, Vincenzo Galilei, who advocated a twelve-tone scale based on what became known as *equal temperament*. In this system, every neighboring pair of notes is separated by the same interval, or ratio of frequencies. With twelve semitones, or half-steps, the width of each successive interval increases by a factor of $2^{1/12}$, or 1.059463. For example, think about the scale that starts from the A above middle C, which has a frequency of 440 hertz (cycles per second) in modern orchestral tuning. The next note up is A sharp, with a frequency of 440×1.059463, or about 466.2 hertz. Twelve intervals up from the starting note brings us to the octave with a frequency of $440 \times 1.059463^{12} = 880$ hertz, or double the starting frequency.

Determined in this way, none of the frequencies of twelve-tone equal temperament (12-TET) exactly matches that of the corresponding notes in just intonation, except at the tonic and octave, although fourths and fifths are so close as to be almost indistinguishable. Equal temperament is a compromise: it isn't as pure sounding as just intonation, but it has the huge advantage of enabling music to be played that is acceptably harmonious in any key without the need for retuning. It made keyboard instruments, such as the piano, practical and musically flexible and opened up broad new horizons in composition and orchestration.

The 12-TET is what's used almost universally in Western music today. But in other parts of the world, different systems of tuning have evolved, which is partly why the music of the East and Middle East has, to our Western ears, an exotic sound to it. Arabic music, for instance, is based on 24-TET, so that it makes liberal use of quarter tones. However, only a fraction of the twenty-four tones appear in any given performance, and these are determined by the *maqam*, or melody type, being used—comparable to how in Western music generally only seven out of the twelve tones appear, which are determined by the key. As in Indian raga and other traditional non-Western forms, there are strict rules, even within the most elaborate and protracted improvisations, that govern the choice of notes and their relationship, together with the patterns of these notes and the progression of the melody.

From an early age, our brains become accustomed to the music that's pervasive around us, just as they adapt to the local language, the tastes of our home food, and the ways of the people with whom we grow up. Music from other cultures may sound unusual and surprising, yet, for the most part, it is still pleasing to the ear. The different scales, intervals, rhythms, and structures of musical pieces from other parts of the world may take some getting used to, but we almost always recognize them as being musical. This is because they too are based on acoustic patterns that can be reduced to relatively simple mathematical relationships that govern such elements as melody, harmony, and tempo.

Whether the concept of music is universal is debatable. Even in the West, there have been many auditory explorations and developments, especially over the past century or so, that test the boundaries of what might be called musical. These include atonal music, which lacks the usual tonal center, and experimental music, which intentionally breaks the customary rules of composition, tunings, and instrumentation. A pioneer of the latter was American composer and philosopher John Cage, whose *4′33″* is a three-movement piece in which the performer (such as a pianist) or performers (up to a full orchestra) are instructed to play nothing throughout. The only sounds heard by the audience are whatever other sounds happen to be going on at the time—someone coughing, the creak of a chair, noises from outside. The inspiration for it came from a visit by Cage to Harvard University's anechoic chamber, a completely echoless room, after which Cage was moved to write, "There is no such thing as empty space or empty time. There is always something to hear or something to see. In fact, try as we might to make a silence, we cannot." Cage intended the piece to be taken seriously, but, inevitably perhaps, others saw a lighter side to it. In his essay "Nothing," Martin Gardner wrote, "I have not heard 4′33″ performed, but friends who have tell me it is Cage's finest composition."

However we choose to define it, music isn't exclusive to humans. Many other species make sounds that we often interpret as being musical, prominent among them being birds and whales. Masters of tuneful performances in the animal world are songbirds, of which more than four thousand species are known, including such families as the larks, warblers, thrushes, and mockingbirds. The songs are normally sung by the male, either to attract a mate or to proclaim his territory or, very often, both. Male sedge warblers, which winter in the Sahara before returning to Europe in the spring, a few days ahead of the females, sing both day and night, since their prospective partners may arrive in either, staking out and defending their territories at the same time, and then abruptly fall silent after they find a mate. Each species has a particular song, which is unvarying, although individuals can distinguish each other's voice prints, just as human voices sound different to us even if they're singing the same tune. The individuals of some species, such as chaffinches, have a repertoire of fixed phrases. If one chaffinch sings a particular phrase, his neighbor will answer him back with a similar phrase—a kind of echo—the purpose of which, according to one suggestion, may be that it allows the two birds to judge their distance apart.

Songbirds certainly appear to be tuneful, and composers, including Vivaldi and Beethoven, have sometimes turned to them for inspiration. But whether any of their songs follow the same kind of organizational rules that humans use in their music is unclear. Some similarities are bound to happen just because of the science of acoustics and the way sounds are produced in throats and mouths. For instance, both we and the avians tend, on the whole, to use adjacent notes that aren't widely spaced in pitch and long notes at the end of phrases. The question is whether birds, like us, favor certain relationships between notes—definite scales—and other orderly patterns in their songs. Not much work has been done on this, but one piece of research, which focused on a particularly tuneful bird, the nightingale wren of Costa Rica and southern Mexico, looked for any intervals in its songs that might correspond with diatonic, pentatonic, or chromatic scales. It found that there was no match at all other than what might be expected by chance. This doesn't mean that the songs have no meaning—at least to other birds—just that they don't follow Western musical scales. The fact that we find the sounds both pleasant and patterned suggests that they are music of some type, even if not of our own ilk.

The vocalizations of cetaceans, including whales and dolphins, are vastly more elaborate than anything a bird produces and are used for both communication and echolocation. The songs of the humpback whale, in particular, have been described as the most complex in the animal kingdom, but they're

A humpback whale breaching. (NOAA)

neither musical pieces nor conversations in the conventional sense. Each song is built up from bursts of sound, or "notes," that may last a few seconds, swoop up or down in frequency or stay the same, and range in frequency from the lowest we can hear to somewhat above the highest. Also the volume of the sound may vary over its duration. Several of these notes together constitute a subphrase, lasting maybe ten seconds, and two subphrases combine to make a phrase, which the whale repeats as a "theme" for a few minutes. A group of such themes makes up a song that may go on for a half hour or so and then be repeated over and over again for hours or even days. At any point in time, all the humpbacks in a given region sing the same song but gradually alter small elements of it, in rhythm, pitch, and duration, as the days go by. Populations of whales occupying the same geographical region have similar songs, whereas those living in different parts of the ocean, or different oceans, have totally different songs, though the underlying structure is the same. So far as is known, once a song has evolved, it never goes back to the original pattern. Mathematicians who've applied information theory to the songs see in them a complexity of syntax and hierarchy of structure previously not encountered outside of human language. But whatever the whales are doing, they're not having regular conversations, because the songs, while subtly and continually changing, are too repetitive. Think of them perhaps as being like jazz or the blues, where riffing and improvisation are allowed, even encouraged, but within well-defined guidelines. A clue to the function of whale songs is that they're performed exclusively by males, and the most creative individuals, in terms of coming up with new variations, tend to be most successful in attracting female partners. It's also hard to avoid the suspicion that the whales are having a great time in these collective jam sessions.

To our ears, there is a beauty and otherworldliness to whale songs that has been captured on CDs intended for relaxation and therapy. Part of a whale song, recorded by marine biologist Roger Payne using hydrophones off the coast of Bermuda in the 1970s, was placed on the Golden Disks now heading toward the stars aboard the twin *Voyager* probes. One of those who helped produce the disk, American science writer Timothy Ferris, suggested that whale song might prove more intelligible to smart aliens than it is to us, so a lengthy piece of it was included and overlapped with greetings in various human languages. As Ferris pointed out, from an alien's point of view, "It doesn't interfere with the greetings, and if you are interested in the whale song, you can extract it."

Music, like love and life, is hard to define. We might say we know it when we hear it, so that the definition becomes one based on personal or collective taste—a purely subjective thing. No one would seriously argue that compositions by Beethoven or the Beatles are not musical. But what of birdsong? What of some of the artistic output of avant-garde sonic artists, like John Cage and Harry Partch, the latter who built instruments to challenge the orthodoxy of modern Western scales and harmonies? If we want an objective definition of music, we must turn to the science of acoustics and the laws of mathematics and ultimately reduce sounds, and combinations of sounds, to numbers. Again, how we choose to do this is up to us, but whatever we choose, it will involve a combination of at least some of the elements without which music is impossible: melody, harmony, rhythm, tempo, timbre, and perhaps others. Once a set of criteria was selected and programmed into a computer, it would be possible to analyze any sound and decide if, according to the chosen rules, it qualified as being music. The criteria could be made as inclusive or exclusive as we wanted, depending on how wide the net was to be cast, but they couldn't be so loose as to include all sounds, or even all regular sounds. The breaking of waves on the shore is a pleasant, soothing sound and has a regular tempo, but most people would probably agree it could hardly be called musical.

Behind all music as we normally understand it is some kind of intelligence. It's possible to imagine a natural system that might be able to produce truly musical passages, in the same way that certain things in nature manifest beautiful spatial forms such as a Fibonacci spiral. But nothing like that has been found to date. So far as we know, it seems that a form of brain, whether it be human, whale, bird, or computer, is needed to construct the type of sonic patterns necessary to qualify as music. Because music is fundamentally mathematical, and because mathematics, to the best of our knowledge, is universal, it seems most likely that if other intelligent species have evolved across the galaxy and beyond, then they too might have come up with music of some form. The variety is likely to be immense, just as it is on earth. Think of a spectrum that embraces Gregorian chant, flamenco, bluegrass, gamelan, Noh, fusion, psychedelic rock, romantic classical, and all of the other genres of music from around the world and across the ages. Now add in the possibility of new genres of which human minds have never conceived, and the extent of what might constitute music across the cosmos becomes apparent. What's more, our appreciation of music is limited by our anatomy, especially the frequency range to which our ears are sensitive, roughly between 20 and 20,000 hertz. Other animals can hear sounds well outside this range: down to 16 hertz in the case of elephants and up to about 200,000 hertz in some types of bat. In theory, there's no limit to the types of sounds alien anatomies might be capable of handling, in terms of frequency, amplitude, ability to discern differences in pitch, tempo, and the like, or any other physical parameter. Possibly the processing power of some extraterrestrials will be vastly greater than that of our own brains, or that of our fastest computers, so that they can appreciate some complex sounds as being musical that would, in a sense, go over our heads.

As for the music that has been included on the *Voyager* Golden Disks, now on their endless journeys through interstellar space, there's been much discussion about which pieces would most obviously sound musical to alien ears. Some believe it would be the works of Bach, that most mathematical of composers. In fact, of the twenty-seven selections of music contained on the disks, lasting a total of ninety minutes, three are by Bach—extracts from the Brandenburg Concerto no. 2 in F, the "Gavotte en Rondeau" from the Partitia no. 3 in E Major for Violin, and the Prelude and Fugue no. 1 in C Major from *The Well-Tempered Clavier*, Part 2. The Bach contributions last twelve minutes and twenty-three seconds, or roughly one-seventh of the playing time of the whole record, reflecting the belief of those who assembled the collection that the highly structured nature of Bach's pieces, including his clever and complex use of counterpoint to interweave multiple melodic lines, would appeal to both the intellect and the aesthetics of any advanced beings who came across the spacecraft.

Scientists and writers alike have pondered what extraterrestrial music might be like. In the movie *Close Encounters of the Third Kind*, the aliens played a five-note sequence from a major scale as a greeting: "re mi do [down octave] do sol." Perhaps, in the story, they did that because they would have been listening to our music and wanted it to sound familiar. Or perhaps other races across the galaxy will come up with the same musical scales as we did, because they're the simplest mathematically and the best ones from which to make attractive melodies and harmonies, whether you grew up on earth or the fourth planet of a star forty thousand light-years away. If mathematics is universal, then so, with many variations, may be the fundamentals of music, including similar scales and methods of tuning. There's a certain inevitability about the development of equal temperament, for example, that may be repeated wherever intelligent beings want to be able to play a variety of different instruments and harmonize them in many different keys.

If or when humans finally make contact with another intelligence among the stars, there's the chance it will be through music. It isn't a new idea. In the seventeenth century, English clergyman Francis Godwin, bishop of Hereford, wrote a book called *The Man in the Moone* (published posthumously in 1638) in which his intrepid astronaut, Domingo Gonsales, encounters a race of Lunarians who communicate via a musical language. Godwin's idea built on a description of the spoken Chinese language, with its tonal sounds, by Jesuit missionaries who had recently returned to Europe. In Godwin's tale, the Lunarians used different notes to represent the letters of their alphabet.

In the 1960s, German radio astronomer Sebastian von Hoerner, who wrote extensively on SETI (the search for extraterrestrial intelligence), argued in favor of music as a medium of choice for interstellar communication. And it might very well be, he suggested, that alien music would share some features in common with our own. Wherever polyphonic music, in which more than one note is played at the same time, evolved, there would be only a limited number of workable solutions that gave harmonious sounds. To allow for modulations from one key to another, an octave has to be divided into equal parts and the corresponding tones have to be at frequencies that bear certain mathematical ratios to one another. The compromise that's emerged in Western music is the 12-TET scale. That scale, said Hoerner, might crop up in the music of other worlds, as might a couple of other scales that offer good compromises for polyphony: the five-tone scale and the thirty-one-tone scale. The latter was written about in the seventeenth century by a number of scholars, including astronomer Christiaan Huygens, and might be the scale of choice among beings whose auditory systems are more sensitive than our own. Aliens whose biology made them less adept at differentiating between closely spaced pitches would perhaps be more likely to use 5-TET.

It's often assumed that the first message we receive from "out there" will be scientific or mathematical in content. But what better way to extend a greeting than by sending a really good piece of music, not just one that has a logical basis but one that is full of the passion and emotions of its creators?

7

PRIME MYSTERIES

> Mathematicians have tried in vain to this day to discover some order in the sequence of prime numbers, and we have reason to believe that it is a mystery into which the human mind will never penetrate.
>
> —LEONHARD EULER

> The greatest problem for mathematicians now is probably the Riemann Hypothesis.
>
> —ANDREW WILES

A prime number is just a natural number that can be divided, without a remainder, by only itself and 1. This might not seem like a particularly special quality, but prime numbers occupy a position of central importance in math. It isn't an exaggeration to say that some of the greatest unsolved mysteries in the subject involve primes and that, on a practical level, these numbers play an important part in our daily lives. Whenever you use a bank card, for instance, the bank's computer checks you're the owner by using an algorithm that cracks a very big number into a unique product of two known primes. Much of our financial security depends ultimately on these numerical oddballs.

The first few primes are 2, 3, 5, 7, 11, 13, 17, 19, 23, and 29. All numbers that aren't prime are said to be composite. The number 1 itself isn't considered a prime—although it could be—because if it were, it would complicate some useful theorems, including one that's so important it's called the fundamental theorem of arithmetic. This states that every number can be written in exactly one way (ignoring rearrangements) as a product of one or more primes—for instance, $10 = 2 \times 5$ and $12 = 2 \times 2 \times 3$. If 1 were allowed to be prime, then there would be an infinite number of ways of writing such a product because it could include any number of 1s multiplied together.

In nature prime numbers crop up in a very unexpected and surprising way. One species of cicada, *Magicicada septendecim*, has a seventeen-year life cycle. All the individuals of this species remain in their larval phase for exactly seventeen years before the whole mass of them emerges at once as adults to mate. Another species, *M. tredecim*, is extremely similar but has a thirteen-year life cycle. There are various theories about why these cicadas evolved specific prime-number life cycles. The most popular is that there existed a predator, which would also emerge in a regular number of years. If both cicadas and predators reached maturity in the same year, that brood of cicadas would very likely have been wiped out. Survival for the cicadas hinged on evolving a life cycle that overlapped as little as possible with the predators. If, for instance, a species had a fifteen-year life

Cicada. (MEGAN SEARING YOUNG)

cycle, the predator could surface every three or five years and kill the cicadas every time they emerged, or surface every six or ten years and kill the cicadas every second time the cicadas emerged, so that the species would rapidly go extinct. However, if a cicada has a seventeen-year life cycle, then if the predator had a life cycle of fewer than seventeen years (which is likely, as evidence suggests the hypothetical predator had a shorter life cycle than the cicadas), then the predator would emerge unsuccessfully sixteen consecutive times and likely die, from starvation, as a result. These predators would have long since disappeared, leaving behind the cicadas we now see with their prime-numbered life cycles.

It's known that there must be an infinite number of primes or—what amounts to the same thing—that there's no largest prime. Euclid proved this more than two thousand years ago. A similar and simple proof runs as follows: Suppose that the number of primes is *not* infinite. Then we would be able to multiply them all together: $2 \times 3 \times 5 \times 7$ and so on, all the way up to the biggest one on the list. Let's call the gigantic product that we would get P and now add 1 to it. There are only two possibilities: either $P + 1$ is prime, or it's divisible by a smaller prime. But if we divide $P + 1$ by any prime on our list of what are supposedly all the primes, there would always be a 1 left over, forcing us to conclude that $P + 1$ must also be prime itself or have a prime factor not on the list. Starting from the assumption that there's a largest prime number, we're led to a contradiction. In logic and math, this is what's called reductio ad absurdum, that is, disproving an argument by showing it has absurd consequences. The starting assumption must be wrong, and therefore its opposite must be true: there are infinitely many primes, a result known as Euclid's theorem.

Mathematicians in ancient times had no easy way to calculate large primes. They would certainly have known in classical Greece that 127 was prime because it's implied by a result in Euclid's *Elements*, and perhaps they knew of others out to values of a few hundred or a few thousand. Significantly larger primes were found in the Renaissance era, including the number 524,287, found by Pietro Cataldi, a prominent prime hunter from Bologna. The search for new primes began to center around numbers of the form $2^n - 1$, where n is an integer, which are now known as Mersenne numbers, after seventeenth-century French monk Marin Mersenne, who devoted a lot of time to studying them. Mersenne numbers are useful "prime suspects" because, selected at random, they're much more likely to be prime than are randomly selected odd numbers of similar size (though not all Mersenne numbers are prime). The first few Mersenne primes (Mersenne numbers that are prime) are 3, 7, 31, and 127. Cataldi's big prime is the 19th Mersenne number (M_{19}) and the 7th Mersenne prime. It took almost a century and a half for a larger prime to be found, by Swiss mathematician Leonhard Euler in 1732. Almost a century and a half after that, in 1876, the record holder became Édouard Lucas, who showed that the 127th Mersenne number (M_{127}), with a value of roughly 1.7 trillion trillion trillion, is also prime.

While many Mersenne numbers are indeed prime, Mersenne himself made a few errors in determining primality, such as believing that M_{67} was prime. The factors were first found in 1903 by Frank Nelson Cole. On October 31, he was invited to give an hourlong talk at the American Mathematical Society. He walked up to the blackboard, without saying a single word, calculated $2^{67} - 1$ by hand, and then calculated 139,707,721 × 761,838,257,287, showing they are the same, before returning to his seat to a standing ovation. He claimed that it had taken him "three years of Sundays" to find the factors of $2^{67} - 1$.

Since 1951 the search for new prime numbers has exclusively involved computers and progressively faster algorithms for seeking out larger and larger Mersenne primes. At the time of writing, the largest known is $M_{74,207,281}$, which has 22,338,618 digits. It was found on September 17, 2015, by Curtis Cooper at the University of Central Missouri as part of the Great

Internet Mersenne Prime Search, a collaborative, distributed computing project of volunteers that has calculated the 15 largest Mersenne primes in the twenty-odd years it's been running. As has become customary, its discoverers marked the event by uncorking a celebratory bottle of champagne.

We know, then, what a prime number is and that the primes go on forever. We know that they can be useful to us in the modern world and that they also crop up in nature. But there are many things we don't know about primes, including whether certain well-known hypotheses are true. One of the more famous of these is the Goldbach conjecture, named after German mathematician Christian Goldbach. This states that every even number greater than 2 can be written as the sum of two primes. For small even numbers, $4 = 2 + 2$, $6 = 3 + 3$, $8 = 3 + 5$, $10 = 3 + 7$, and so on, this is easy to verify. Much larger numbers have been checked using computers, and the rule has never been found to fail. However, no one knows if it's true in the general case.

Another unsolved conjecture has to do with pairs of prime numbers that differ by just 2, such as 3 and 5 as well as 11 and 13. These are called twin primes, and the so-called twin-prime conjecture is that there are infinitely many of them. To date, though, no one has been able to show that this is true beyond doubt.

Perhaps the greatest mystery of all to do with primes concerns their distribution. Among small natural numbers, primes are very common, but they become sparser and sparser as the size of the numbers grows. Mathematicians are interested in the rate at which the thinning out happens and how much we can know about the frequency of prime numbers. Their occurrence doesn't follow any strict regular pattern, but that isn't to say they just spring up willy-nilly. In *The Book of Prime Number Records*, Paulo Ribenboim put it this way: "It [is] possible to predict with rather good accuracy the number of primes smaller than N (especially when N is large); on the other hand, the distribution of primes in short intervals shows a kind of built-in randomness. This combination of 'randomness' and 'predictability' yields at the same time an orderly arrangement and an element of surprise in the distribution of primes."

Countless mathematicians have commented on the enigmatic nature of prime numbers. They're the simplest of things to describe—so simple that children in elementary school are taught what they are and are often asked to name the first few of them or say whether a number is prime. Agnijo himself became fascinated with primes at a very early age and with some of the unsolved problems that surround them. In time, this led to his fascination with other great mysteries of number theory.

Primes are also much like the atoms of the numerical universe, from which all other natural numbers are built. You would think there would be every reason to hope and suppose that they obeyed strict laws and that it should be easy to predict where the next one occurred along the number line. Yet these most elemental of mathematical building blocks are shockingly unruly and capricious in their behavior. It's this tension between expectation and reality, and the strong suspicion that some organizing principles of great importance lie just beyond our grasp, that has fascinated mathematicians since antiquity.

Looked at individually, or in small groups, the primes do indeed appear lawless. But viewed en masse, like shoals of fish or murmurations of starlings, a previously hidden level of organization emerges. One of the most curious discoveries about them happened by accident. While he sat listening to a dull lecture in 1963, Polish mathematician Stanislaw Ulam started doodling on a sheet of paper. He wrote down a square spiral of numbers, starting with 1 in the center and gradually working his way outward along a rectangular grid. He then circled all the primes and noticed something surprising. Along certain diagonals of the spiral, as well as some horizontal and vertical alignments, prime numbers were unusually dense. Larger Ulam spirals, produced using computers and containing tens of thousands of numbers, continue to show these patterns. In fact, it seems that they stretch out as far as we care to calculate.

Some of the prominent lines in the spiral correspond with certain formulas in algebra that are known to generate a lot of prime numbers. The best known of these was discovered by Leonhard Euler and is named after him. Euler's "prime-generating polynomial," $n^2 + n + 41$, spits out primes for every single value of n from 0 up to 39. For instance, for $n = 0, 1, 2, 3, 4,$ and 5, it outputs 41, 43, 47, 53, 61, and 71, respectively. For $n = 40$, it gives the (nonprime) square number 41^2 but continues to yield a very high frequency of primes as n gets larger. There are other, similar, formulas that, for some reason that isn't clear, have this peculiar ability to spawn primes at a great rate. Mathematicians continue to discuss the significance of the patterns in the Ulam spiral and their connection with unsolved problems such as Goldbach's conjecture, the twin-primes conjecture, and the hypothesis known as Legendre's conjecture that there's always at least one prime between consecutive perfect squares.

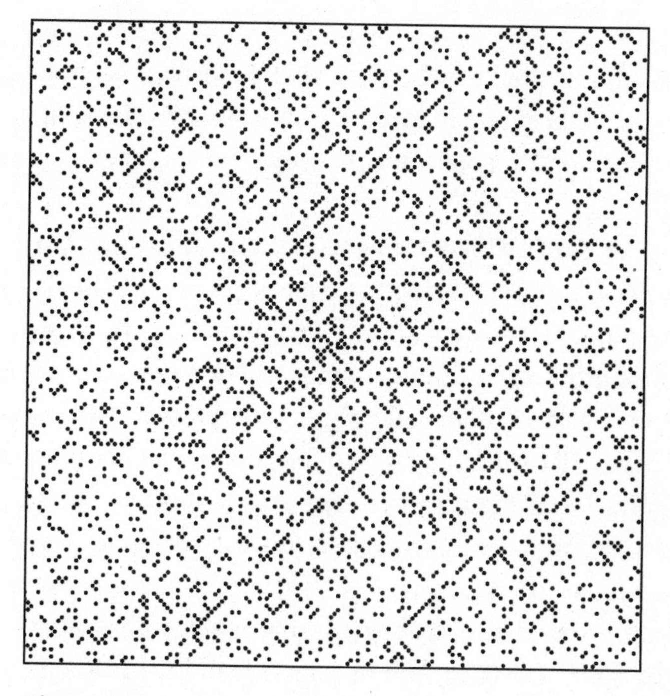

Ulam spiral. (AGNIJO BANERJEE)

What the spiral makes graphically clear, however, is that there *are* patterns and that despite appearing haphazard in their distribution, primes follow some overarching rules that govern their behavior in large groups.

The best theorem we have about how the prime numbers are distributed is called, unsurprisingly, the prime-number theorem and is widely regarded as one of the greatest achievements in number theory. In a nutshell, it says that for any number N that is large enough, the number of primes less than N is roughly equal to N divided by the natural logarithm of N. (The natural log of a number is just the power to which the number e, which is equal to 2.718 . . . , has to be raised to equal the number.) This formula doesn't tell us where the next prime will lie, but it does give a pretty accurate indication of how many primes there are within a given number interval, providing the interval is large enough.

Unlike Euclid's theorem of the infinity of primes, which, as we've seen, can be explained in a few lines of plain English, the prime-number theorem took a century of effort to prove. It was first mooted by the German Carl Gauss, as a teenager, in 1792 or '93, and independently by Frenchman Adrien-Marie Legendre a few years later. Of course, mathematicians had long recognized that the gaps between primes tend to get wider as the size of numbers increases, but it was the publication of extended tables of primes and of longer, more accurate tables of logarithms, in the second half of the eighteenth century, that helped spur efforts to find specific formulas to describe this thinning out. Gauss and Legendre spotted that a one-over-log type of function was at work. Further important progress toward refining the distribution formula was made by Russian mathematician Pafnuty Chebyshev between 1848 and 1850. But the biggest breakthrough of all came through the efforts of German Bernhard Riemann who, in 1859, published an eight-page memoir, his only writing on the subject, titled *On the Number of Primes Less Than a Given Magnitude*. In it he put forward a suggestion, subsequently called the Riemann hypothesis, that has teased and tormented mathematicians ever since in their attempts to prove it. David Hilbert reputedly said that the first thing he would ask after waking from a sleep lasting a thousand years would be "Is the Riemann hypothesis established yet?" In his book on the theory behind Riemann's suggestion, American mathematician H. M. Edwards wrote: "It is now unquestionably the most celebrated problem in mathematics and it continues to attract the attention of the best mathematicians, not only because it has gone unsolved for so long but also because it appears tantalizingly vulnerable and because its solution would probably bring to light new techniques of far-reaching importance."

To underscore how highly the Riemann hypothesis is regarded, it's one of the seven Millennium Prize Problems identified by the Clay Mathematics Institute of Cambridge, Massachusetts, and for which a $1 million prize is on offer for the first verified solution. It's one of the two that Agnijo would particularly like to solve, the other being the P versus NP conjecture (discussed in Chapter 5). The Riemann hypothesis is also the only one of the Millennium Prize Problems that appears in a list of twenty-three major unsolved problems discussed by David Hilbert in an address he gave, on August 8, 1900, to the International Congress of Mathematicians in Paris.

To the question of how the primes are distributed, Riemann brought to bear the methods of a newly developed branch of math known as complex analysis. As the name suggests, this has to do with all the different ways of working with complex numbers—numbers that contain a real part and an "imaginary" part, such as $5 - 3i$, where i is the square root of -1. At the core of complex analysis is the study of complex functions, which are just rules for turning one set of complex numbers into another. Back in 1732, the prodigious and astonishingly creative Swiss mathematician Leonhard Euler, whose collected works run to more than thirty-one thousand pages, defined a previously unknown beast of the mathematical world called the zeta function. It's a type of infinite series—an infinitely long sum of terms that may or may not converge to a finite value depending on what numbers are fed into it. Under certain circumstances, the zeta function reduces to a series similar to the harmonic series $1 + \frac{1}{2} + \frac{1}{3} + \frac{1}{4} + \ldots$, which has been studied since ancient Greek times when Pythagoras and his followers were obsessed with understanding the universe in terms of numbers and musical harmony. Riemann took Euler's zeta function and extended it to include complex numbers, which is why the complex zeta function is also known as the Riemann zeta function.

In his famous memoir of 1859, Riemann put forward what he believed was a better formula for estimating how many primes there are up to a given number. However, this formula depends on knowing for which values the Riemann zeta function is zero. The Riemann zeta function is defined for all complex numbers of the form $x + iy$ except those for which $x = 1$. The function goes to zero for all negative even integers (–2, –4, –6, and so on), but these aren't of interest in tackling the problem of how prime numbers are distributed and so are referred to as "trivial" zeros. Riemann realized that the function also has an infinite number of zeros in a critical strip between $x = 0$ and $x = 1$ and, further, that these "nontrivial" zeros are symmetric with respect to the line $x = \frac{1}{2}$. His famous hypothesis is that all of the nontrivial zeros of the complex zeta function lie, in fact, *exactly* on this line.

If true, the Riemann hypothesis implies that prime numbers are distributed as regularly as they possibly can be within the ultimate limits imposed by the prime-number theorem.

In other words, granted that there's a certain amount of "noise" or "chaos" that introduces uncertainty into where prime numbers spring up, the Riemann hypothesis says that noise is extremely well controlled—that the apparent indiscipline of primes is, behind the scenes, highly choreographed. Another way to think of this is in terms of rolling a multi-sided die that has a probability of $1/\log n$ of coming up prime. Suppose for each integer n that is greater than or equal to 2, you roll the die n times. In an ideal world, the expected number of primes would be $n/\log n$. But the world not being ideal, there's always some variation—a margin of error—around the expected value. The size of this error is given by what's commonly called the law of averages (or large numbers). What the Riemann hypothesis claims is that the deviation of the distribution of primes from $n/\log n$ is no greater than what the law of averages predicts.

There's plenty of strong evidence to suggest that the Riemann hypothesis is true. Riemann checked the first few nontrivial zeros himself to make sure they obeyed his rule, and, with one of the earliest computers, Alan Turing ran the calculation out to the first thousand. In 1986 came verification that the first billion and a half nontrivial zeros of the Riemann zeta function sit right on the critical line where the real part of the function equals ½. Much earlier, in 1915, G. H. Hardy proved that there's an infinite number of nontrivial zeros (though not necessarily all of them) on this line. In 1989 American mathematician Brian Conrey showed that the number of zeros on the line had to be more than two-fifths of the entire population of zeros in the critical strip. Six years later, after several years of running ZetaGrid, a distributed computing project, the first

100 billion zeros of the Riemann function were found to fall precisely on the critical line, with no exceptions.

It would be perverse to suspect that the Riemann hypothesis was wrong given every indication that it's right. However, in mathematics, belief and persuasive evidence are a world away from proof. Without proof, any results, however useful, that take for granted the mere suggestion of a theoretician, even one as eminent as Bernhard Riemann, are like a house resting on sand. While there remains the possibility that a single nontrivial zero lies elsewhere in the critical strip other than on the line $x = ½$, this wonderful notion of Riemann's really has no more weight than wishful thinking.

The importance of proving (or disproving) it, however, goes well beyond the bounds of number theory or of mathematics as a whole. The Riemann hypothesis, it turns out, has a subtle but direct connection with the subatomic universe. One day in April 1972, at the Institute for Advanced Studies, in Princeton, New Jersey, mathematicians Hugh Montgomery and Atle Selberg were chatting about Montgomery's recent discovery to do with the spacing of nontrivial zeros on the critical line. Later, in the cafeteria, Montgomery was introduced to Freeman Dyson, who was a professor in the School of Natural Sciences. As soon as Montgomery mentioned the subject of his work on the zeros, Dyson realized that the math was identical to that of a theory he had investigated in the 1960s. So-called random matrix theory can be used to carry out calculations on the energy levels of particles inside heavy atomic nuclei. Dyson recalled his surprise at seeing the same equations pop up in a field to do with the distribution of prime numbers: "His result was the same as mine. They were coming from completely different directions and you get the same answer. It shows that there is a lot there that we don't understand, and when we do understand it, it will probably be obvious. But at the moment, it is just a miracle."

Often aspects of math, such as the Riemann hypothesis, seem completely abstract and of no interest except as some elaborate intellectual exercise. Yet here's an example—and they're not as uncommon as it may appear—of a direct connection between seemingly pure mathematics and the physical universe on a fundamental level.

More than 150 years have passed since Riemann announced his hypothesis to the world, and the absence of a proof has become like a gaping hole at the heart of mathematics. Perhaps the ideas needed to solve it are so advanced or radical that they're beyond the scope of our present understanding. If so, the very pursuit of a proof may help develop powerful new mathematical techniques. If a proof finally does come, its importance to math will be hard to overstate because of the foundational role that prime numbers play in the number system and the relationship they have to hugely diverse areas of the subject. Hundreds of theorems will stand or fall depending on whether the Riemann hypothesis is found to be true or false. If true, more questions will be asked, including *why* primes lie at such a delicate balance point between randomness and order. If false, then all these theorems would collapse and a devastating earthquake would shake mathematics to its core.

No one is expecting the Riemann hypothesis to be proved or disproved anytime soon. Yet sometimes proofs appear in mathematics suddenly and without warning. This was the case with the magnificent vindication of Fermat's last theorem by Andrew

Wiles. It also happened more recently in connection with a discovery related to the twin-primes conjecture—the idea, widely believed to be true, that there are infinitely many twin-prime pairs. In 1849 French mathematician Alphonse de Polignac went further and proposed that there were infinitely many prime pairs for any possible finite gap, not just 2. Little progress had been made in proving these suggestions until, out of the blue, in 2013, a middle-aged lecturer at the University of New Hampshire named Yitang Zhang, unknown in the wider mathematical community, published a paper with a startling result. Zhang had managed to show that there's a number, *N*, smaller than 70 million, such that there are infinitely many pairs of primes that differ by *N*. What this means is that no matter how far we wander into the remote lands of vast and ever-vaster prime numbers, and no matter how sparse the primes in general become, we'll always continue to find prime pairs that differ by fewer than 70 million. There's every reason to believe that this number can be greatly reduced and to hope, in a wider sense, that some remarkable breakthroughs in prime-number research are on the horizon.

While prime numbers are quite simple to understand, they form mysterious patterns that we still haven't properly explained. Is every even number the sum of two primes? Are there infinitely many pairs of primes differing by 2? No one knows for sure, although many think that we're close to an answer. Prime numbers also seem to be fundamental to practically all of mathematics—and, perhaps, to the physical universe itself.

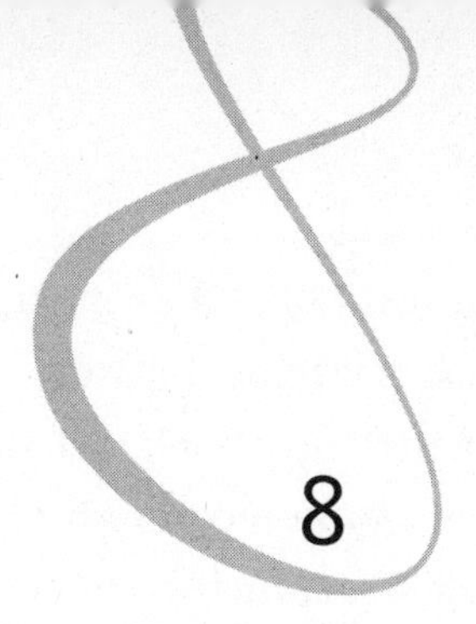

8

CAN CHESS BE SOLVED?

> Chess is a unique cognitive nexus, a place where art and science come together in the human mind and are then refined and improved by experience.
>
> —GARRY KASPAROV

Imagine an incredibly powerful computer that could always figure out the best move in any given possible chess position. "Best move" means the one that leads most quickly to winning or, at the very least, not losing—in other words, the optimal outcome for the player. Now, suppose that this computer played against another that was identical to it in every respect. Which computer would win, or would it always be a draw? We've solved so many monumental problems in math, you would think an old game like chess, with easy-to-learn rules, would hold no challenges to theoreticians armed with the latest computing technology. But nothing could be further from the truth.

The first chess-playing machine, known as the Turk, was actually a fake—although it managed to fool a lot of people between when it was first unveiled by its Hungarian inventor,

Wolfgang von Kempelen, in 1770 and its destruction by fire in 1854. Among those who saw it in action were Napoléon Bonaparte (no mean mathematician himself), Benjamin Franklin, and one of the pioneers of modern computation, Charles Babbage. Behind a large wooden cabinet were the head and upper body of a life-size mannequin dressed in impressive Ottoman robes and a turban. Three doors at the front of the cabinet could be opened to reveal an intricate mechanism and other components, while three doors at the back could also be opened, one at a time, to let spectators see through to the other side. What they didn't see, however, was the expert human chess player who sat on a seat that could be slid from one side of the cabinet to the other as the doors were successively opened and closed. The concealed occupant determined the moves made in reply to whoever was challenging the machine and then operated the Turk's arm and hand to move chess pieces on a board visible to the audience via linkages connected to a pegboard chessboard inside the cabinet. Ingenious and exquisitely made though von Kempelen's automaton was, it relied wholly on human brainpower to overcome its opponents.

No mechanical wizardry—no symphony of cogs and gears, levers and linkages—could work fast enough to play even a modest game of chess, such is the complexity of the game. Hopes for a true chess-playing machine had to await the development of the electronic computer after World War II. Pioneers of computation, such as Alan Turing, John von Neumann, and Claude Shannon, were interested in chess as a means of testing out early ideas in artificial intelligence. In a seminal paper on the subject, in 1950, Shannon wrote, "Although of no practical importance, the question is of theoretical interest, and it is hoped that . . . this problem will act as a wedge in attacking other problems—of greater significance." A couple of years later, Dietrich Prinz, a colleague of Turing's, ran the first chess-playing program on the new Ferranti Mark I computer at Manchester University. Memory and processing limitations meant that it could solve only "mate-in-two" problems—in other words, finding the best move when checkmate is two moves away. A reduced version of chess, using a six-by-six board without bishops, was programmed to run on the MANIAC I computer at Los Alamos Laboratory in 1956. The computer played three of these "anticlerical" games: the first against itself; the second against a strong human player handicapped by not having a queen, which the computer lost; and the third against a novice who had only just learned the rules. In this final game, the computer won, albeit against a feeble opponent, thus marking the first victory of machine over human.

In 1958 a researcher at IBM, Alex Bernstein, wrote the first program that could play games of standard chess on the firm's 704 mainframe—the computer on which both FORTRAN and LISP were developed and on which speech synthesis was first achieved. The scene in the film *2001: A Space Odyssey* in which the HAL 9000 computer's awareness gradually degrades as Dave Bowman disconnects its cognitive circuits, was inspired by Arthur C. Clarke having seen the 704's efforts at speech synthesis a few years earlier. Earlier in the movie, HAL easily beats astronaut Frank Poole at chess. Director Stanley Kubrick was a passionate chess player, so, not surprisingly, the moves shown in the HAL-Poole encounter are from an actual game, between A. Roesch and W. Schlage in Hamburg in 1910.

The challenge facing all chess-playing machines is the immense complexity of the game in terms of strategy and possible moves. All told, there are about 10^{46} possible positions and at least 10^{120} unique games of chess, the latter known as Shannon's number, after Claude Shannon, who stated it in his 1950 article "Programming a Computer for Playing Chess." At the start, things are fairly simple, with just twenty possible moves for White—sixteen involving pawns, only three of which are common, and four involving knights, only one of which is common. But the number of possibilities rapidly grows as the game progresses and other pieces, including bishops, rooks, queen, and king, join in the action. After each player has made one move, there are 400 different possible positions, after two moves 72,084 positions, after three moves more than 9 million positions, and after the fourth move more than 288 billion different possible positions. This is roughly one for every star in our galaxy, while the total number of chess games is far greater than the number of fundamental particles in the universe.

In the early days of computer chess, the relatively primitive hardware available was a serious handicap. But the basic programming approach to playing a strong game had already been figured out in the 1950s by Hungarian-American mathematician John von Neumann. The MiniMax algorithm is so-called because it strives to minimize the opponent's score while maximizing its own score. By the end of the decade, it had been combined with another approach, known as alpha-beta pruning, which uses rules of thumb, or heuristics, distilled from the playing strategy of top human players, to weed out bad moves early on so that the computer doesn't waste time going down fruitless branches of its search tree. This is not the same as a computer learning from its mistakes—that came later—but rather an attempt to program in some good tips and move combinations used by grandmasters in the past.

As computers became more powerful in the 1970s and 1980s, they were able to run programs that searched both deeper and smarter. In 1978 a computer won a game against a human master for the first time. The same decade saw the start of the World Computer Chess Championships. One of the authors (David), while employed as applications software manager at the supercomputer maker Cray Research in Minneapolis, worked with Robert Hyatt of the University of Alabama at Birmingham to optimize Hyatt's chess program, Blitz, to run on the Cray-1—then the fastest computer on earth. In 1981 Cray Blitz became the first computer to achieve a master rating after it won the Mississippi State Championship 5–0, and in 1983 it beat its archrival, Bell Labs' Belle, to become the world computer chess champion.

Since that time, progress in computer chess has been dramatic. In 1997 human world chess champion Garry Kasparov lost in a five-game tournament to IBM's Deep Blue, and the last time a human beat the strongest computer on the planet was in 2005. The top computers are now so far beyond the rating ever achieved by a human that it's safe to say that no one will ever defeat the best computer chess players again. At the time of writing, the highest rating (based mainly on tournament wins or losses against other strong players) ever achieved by a carbon-based life-form is 2882, in May 2014, by the current human champion, Norwegian Magnus Carlsen. This is

Agnijo's chessboard at home. The position shown occurred between Deep Blue (*White*) and Garry Kasparov (*Black*) in 1996 in the game in which, for the first time, a computer defeated the human world champion. (AGNIJO BANERJEE)

presently outstripped by at least fifty of the best computer programs, including Stockfish, which has the highest rating ever achieved, by human or machine, of 3394.

Still, for all the prowess of today's high-speed chess-playing systems, the question remains: Is chess solvable? Put another way, can the outcome be known even before play starts? There are many simpler games where the answer to this question is yes. One of the simplest and best known is tic-tac-toe, or noughts and crosses. Analyzing tic-tac-toe is quite easy because the game has to end in at most nine turns, and much of the time a player is forced to play a certain square to stop their opponent from winning. Any games between players who have figured out the strategy will always end in a draw. The fact that tic-tac-toe involves just a three-by-three grid helps makes solving it easy. But games don't need a big board in order to be complex.

Many people, at one time or another, have played the game of dots and boxes, where you start with a square grid of points, and each player in turn draws a line between any two of them. The person who joins up the fourth side of a box wins the box, puts their initial in it, and then connects another pair of dots as part of the same turn. If that results in completing a second box, they join another pair of dots, and so on.

The smallest board size on which the game is interesting is three-by-three. Although this is the same board size as tic-tac-toe, the amount of strategy involved is already far greater. It's known that in a three-by-three game of dots and boxes, the second player can force a win, but most people don't know the winning strategy, which turns out to be surprisingly complex. The majority of us essentially play at random, trying not to give away any boxes, then taking as many boxes as we can, before giving the opponent as few boxes as possible. For boards that are much bigger than three-by-three, theoreticians have no clue as to who will win at the start. They can also find positions, which turn up frequently in high-level play, where it's guaranteed that every move by a player will cause them to lose the game, but after any such move it's not known *how* the other player can win, even though it's known that they can. This is an example of what's called a nonconstructive proof, that is, a proof that shows that something—such as a winning strategy—exists without giving any hint as to how to achieve that end. These kinds of proof may seem counterintuitive. After all, how can we know for sure that something exists without being able to give an example? Yet they turn up frequently in games such as this. The bottom line is that it may be a simple matter to prove that a certain player can win but be completely unfeasible to know in detail exactly how this win is to be achieved.

With dots and boxes, like tic-tac-toe, all the possible moves are available at the start of the game, and the number of possibilities always goes down as the game progresses. Chess is a much more complex game than dots and boxes, which itself already has huge potential for high-level and grandmaster play. In chess many more moves are available at each turn, the number of possible moves expands rapidly, and games can go on for much longer. In terms of knowing who is going to win, the best we can do at present is solve this problem in the case of some endgames where a small number of pieces are left on the board. Solving chess in its entirety—finding an optimal strategy by which one of the players can always force a win or both can force a draw—seems like a distant pipe dream. Having said this, computers have made spectacular progress in being able to look many moves ahead and select powerful sequences of moves from among billions that are available.

Perhaps even more surprising is the rapid progress that computers have made in another ancient and even more strategically complex game, Go. Played mainly in China, Japan, and South Korea, on a nineteen-by-nineteen board, it has roots that extend back two and half thousand years and is the oldest board game still enjoyed today. In antiquity it was one of the four arts of the Chinese scholar, along with painting, calligraphy, and playing the *qin*, a stringed instrument. The antagonists in Go, Black and White, take turns, but unlike in chess, Black plays first. Each player in turn places a stone of their color on the board and can capture groups of opposing stones, removing them by surrounding them (the name Go comes from the Chinese for "encircling game") with their own stones. As well as these basic rules, there are many others, but more than anything the tactics and strategy involved in Go are fiendishly intricate. Tactics refers to what's happening in a local part of the board where groups of stones vie over life, death, rescue, and capture, whereas strategy takes in the global situation of the game. Compared to chess, Go involves a larger board, many more alternatives to consider per move, and generally longer games. The brute-force methods that give chess computers the edge would take far too long to be applied to Go. They would invariably prove useless against a grandmaster, who can decide among the many move options available using higher-level skills, such as pattern recognition, built up from long experience and at which the human brain is particularly adept. Recognizing certain types of pattern, which may superficially look quite different in different situations, is a much harder challenge for computers than simply calculating at lightning speed. In fact, after computers started beating the strongest human players at chess, Go experts remained confident that it would be a long time until computers reached even a modest amateur level at their own game.

Then, in 2016, Google's program AlphaGo defeated one of the world's best Go players, Lee Sedol, by four games to one. Relying not so much on brute-force methods of looking at many game situations ahead, AlphaGo was designed to play in

A game of Go.

a more humanlike way. It's based around a neural network that simulates how an organic brain tackles problems. Starting out with a huge database of expert games, it was made to play a very large number of games against itself, with the aim of eventually learning how to recognize winning patterns. It uses the smart, heuristic approach of a human player combined with the speed of silicon circuitry to achieve what hadn't been thought possible anytime soon, to become a world-class Go superstar. In 2017 AlphaGo went one better, winning three out of three games against the top-ranked human player, China's Ke Jie.

There seems little doubt that, before long, Go-playing computers will be as unbeatable by their flesh-and-blood creators as chess computers are today. But the question remains: Are games like chess and Go ultimately solvable? In chess, because White always goes first, Black can only react to the threats that White poses. So if ever chess were to be solved—in other words, the best sequence of moves found that White could play in response to whatever the opponent did—it's almost certain that the only possible outcomes would be a win for White or a draw. With Go it is less clear because, unlike chess, Black goes first and White receives a number of points (6.5 under Japanese rules, 7.5 under Chinese rules) as compensation. It may be that this is enough for White to win, or perhaps the first-move advantage for Black is so great that Black still wins. No one knows and, perhaps, no one will.

A surefire way to solve chess would be to draw a tree for all possible positions and then, starting from any position, evaluate every branch by looking at where it ended and then choosing the one that led to the optimal outcome. That's fine, in theory. But given that there are roughly 1,200 trillion trillion trillion trillion trillion trillion trillion trillion possible chess games, the resulting tree would be colossal. Building a computer to hold so much data would be a challenge, given that there are probably fewer than 10^{80} atoms in the entire visible universe, which is forty powers of ten smaller. In practice, a lot of the branches could be pruned at an early stage because many of the possible positions are ridiculous and would never crop up in a real game, even between beginners. But after all the clever trimming had been done, the tree of possible, realistic moves would still be extraordinarily daunting. It would be even more so in the case of Go. This massive complexity has led some to conclude that although mathematics doesn't stand in the way of solving such games, matters of practicality do. When there aren't enough subatomic particles in existence to store the tree of moves, even after extensive pruning, how can a solution be achieved? Perhaps advanced artificial intelligence will come to the rescue, enabling vastly more pruning so that the tree size becomes manageable. Quantum computers, able to search huge numbers of branches simultaneously, might be another option, although, unlike with Shor's algorithm for factoring large numbers, we currently don't have an algorithm for solving these types of problems or even know whether any algorithm exists. Some hope for a possible future solution comes from the fact that the game of checkers, or draughts, was solved, in 2007, after hundreds of computers, working over a period of nearly twenty years, searched through all the combinations of moves that could be played. A game of checkers, it transpires, will always end in a draw if neither player makes a mistake. Whether, with advances in technology and programming, chess will eventually follow, and perhaps Go too, remains to be seen.

What we do know is that games like chess and Go, and simpler ones, including tic-tac-toe and dots and boxes, are "games of perfect information." This means that before a player makes a move, he or she has all the information they need to determine which moves are good or bad. Nothing is hidden from view or uncertain. This means that, in principle, given an unlimited amount of memory and time, they could be solved. But there are other games, such as poker, that lack perfect information. When deciding what to do next, a poker player doesn't know what cards anyone else is holding, even though that's a crucial factor in deciding who wins. In a poker tournament that contains a beginner and an expert, the beginner may get lucky, draw a royal flush, and win a hand. However, on average, the expert's superior knowledge of when to bet or fold will see them win more often and win more money than the beginner over the course of a large number of hands.

Before we could ever say that a game like poker has been solved, we would need to be clear about what "solved" actually means when it comes to games without perfect information. No computer could guarantee a 100 percent win rate in poker—without cheating—as the possibility of a human drawing a royal flush would always exist. What could qualify as solving poker is for a computer to play by a strategy that will, on average, result in the maximum winnings.

Poker is further complicated by the possibility of bluffing and the fact that, in most tournaments, considerably more than two people are playing. In a situation with multiple humans and one computer, it's possible—perhaps even likely—that the human players would gang up in such a way as to put the computer at a disadvantage. If they did, each person might win less than if they played entirely selfishly, but the humans as a whole would win more.

All this said, a program has been developed that can't be beaten, over long periods of play, at a type of poker called heads-up limit Texas Hold 'Em (a two-player game). The new software, announced in 2014, marks the first time an algorithm has been found that effectively solves a complex game in which some information is hidden from the player. This hidden information, together with the luck of the draw, ensures that the program can't win every hand. But on average, and over many hands, there's virtually no chance that a human could ever beat it (just as, for example, a human could practically never beat Stockfish at chess), so that effectively this version has been solved. Not only can the program help human players improve their game, but it's also been suggested that the approach it takes could prove useful in health care and security applications.

It may seem, from the poker example, that all games with imperfect information involve some sort of chance that's outside the control of the players. But that isn't the case. In the familiar game of rock-paper-scissors, all that matters is what each person plays: there's no chance involved that is outside the control of the players. Yet in spite of this, the game has imperfect information. The usual way of playing the game, by two people making hand gestures simultaneously, is effectively no different from what would happen if the players were in separate rooms and wrote down their decisions, each unaware of the other player's choice.

Now, in a game with perfect information, there's always some "pure" strategy—some move or series of moves that results in the most favorable outcome. For example, in chess there's always a best move (or often multiple winning moves), which, if played consistently in the same situation, is the optimal thing to do. In the case of rock-paper-scissors, the exact opposite is true. Adopting a pure strategy by playing, say, rock every time, or a regular pattern of rock, paper, scissors, would be easily beaten. Instead, the best approach is what's known as a mixed strategy, which means that in any position, different actions are taken with different probabilities. Solving a game like rock-paper-scissors or two-player poker is about finding an optimal mixed strategy that guarantees the highest probability of winning. The strategy of "always play rock" would have a 100 percent winning probability if the opponent were foolish enough to always play scissors. On the other hand, given the more likely case that the opponent would quickly respond by always playing paper, the winning probability of "always play rock" would fall to 0 percent. It comes as no surprise that rock-paper-scissors has been solved, and the solution is quite trivial. The optimal strategy is to play rock one-third of the time, paper one-third of the time, and scissors one-third of the time. Counting a draw as half a win, this gives the player a 50 percent minimum win rate, which is the best of all possible strategies. While there's some scope for high-level play, it relies on psychology instead of game theory, exploiting the fact that humans are generally bad at being truly random, as we saw in Chapter 3. In general, in games without perfect information, the optimal strategy is always mixed.

In such games, too, there's a concept known as the Nash equilibrium, named after American mathematician and economist John Nash, who made important contributions to game theory and was the subject of the novel (and subsequent film) *A Beautiful Mind*. In a strong Nash equilibrium, all players have a strategy, and if they deviate from it in any way (assuming no one else does so simultaneously), they'll be worse off than before. There's also another concept, a weak Nash equilibrium, where a player can deviate from the strategy and be neither worse off nor better off than before, but it's impossible to deviate from the strategy and end up better than before. Nash equilibriums play a pivotal role in game theory.

In a game with perfect information, a Nash equilibrium occurs if both sides play the optimal strategy. This can be strong or weak, depending on whether there are multiple optimal strategies. In a game with imperfect information, this is also true. However, it's entirely possible for there to be multiple Nash equilibriums. To determine whether we've found them all, another concept is needed, known as a zero-sum game or constant-sum game.

In a zero-sum game, one person's gains exactly equal the other person's losses. More general is a constant-sum game, in which the total gains made by the players never change. Chess is an example. The players could both draw, earning half a point each, or one could win, with the winner earning one point and the loser earning nothing. By contrast, a game such as soccer isn't a constant-sum game because if the teams draw, each gains one point, but if one team wins, they gain three points and the loser gains nothing. The sum of points can be two or three. Constant-sum games can all be converted into zero-sum games by adding or subtracting points. For example, if half a point were deducted from each player of a chess game, the game would be zero-sum. For this reason, results that apply to zero-sum games generally also apply to constant-sum games.

In any zero- or constant-sum game, the only Nash equilibriums occur when both players play an optimal strategy. However, this result doesn't apply to games that are not constant-sum, which may have many other Nash equilibriums. In games that aren't constant-sum, another issue becomes relevant—that of Pareto efficiency. Any set of strategies is Pareto efficient if it's impossible to change all strategies to make someone better off without making someone else worse off. In a zero-sum game, any set of strategies is Pareto efficient. But in general, this isn't the case. Even Nash equilibriums may not be Pareto efficient, as a puzzle known as the Prisoner's Dilemma makes clear.

Two prisoners have both been convicted separately of a crime that carries a sentence of one year. In addition, however, some witness statements have linked the pair to having jointly taken part in a more major crime, with a sentence of six years. The prisoners are given a choice. They can both either remain silent or privately betray their partner. Neither will be told what the other has done until they receive their sentence. If both betray each other, both will receive four years in prison in total (three years each for the major crime and one year each for the minor crime). If only one betrays the other, the betrayer goes away free and the other prisoner receives a full seven years for both crimes. If both stay silent, both can only be convicted of the minor crime and both serve one year. Surprisingly, it turns out that no matter what the other person does, betraying them is always better than staying silent. The only Nash equilibrium is therefore when both prisoners betray each other and both serve four years. However, this isn't Pareto efficient, as it's better for both of them to stay silent and serve only one year each. The Prisoner's Dilemma can be repeated any number of times, with strategies that depend on what happens in the past—a problem known as the iterated Prisoner's Dilemma, which can become very complicated indeed. The best strategies for the iterated version tend to be those that generally stay silent, provided the other player also does so, but punishes players for betraying them by also betraying. These strategies thereby reap the benefits of the Pareto-efficient outcome against each other, while trying to avoid the worst outcome by opting for the Nash equilibrium if it's clear that the other strategy is betraying it.

Most people prefer that the games they play end in a reasonable amount of time—say, an hour or two—before fatigue, hunger, or boredom sets in. The World Chess Federation puts a time limit, for all its major events, of ninety minutes for the first forty moves and thirty minutes for the rest of the game. However, the longest game on record, between Ivan Nicolic and Goran Arsovic in Belgrade in 1989, lasted more than twenty hours before ending in a draw after 269 moves due to the "50-move" rule. This rule states: "The game may be drawn if each player has made at least the last 50 consecutive moves without the movement of any pawn and without any capture." A draw can also be claimed—again, by the player whose turn it is—if the same position has occurred three times. Assuming that such a claim is made under the 50-move rule, the longest a game can possibly go on is just under 6,000 moves.

Much longer, potentially billions of times longer than the sun will shine, are games that could be played on a chessboard that stretched forever in all directions. So-called infinite chess has the same rules and number of pieces as the common or garden finite variety but uses a board that has no end or edges. Playing it could involve some spectacular moves—a rook might shoot off a trillion places in one direction, and a bishop might hurtle in from the equivalent distance of intergalactic space to capture a pawn in the next. This is not a socially acceptable game for limited beings like ourselves. Yet through the power of math, we can know something about it even if we can never participate. Most important, we can be sure of one very important fact about infinite chess: like it's finite cousin, there is a strategy, which, if adopted, would guarantee a win for one of the players. What is that strategy? Unless we had a computer of infinite speed and memory capacity, there's no way of knowing. But the fact that all forms of chess, and other games of perfect information, finite or infinite, can be solved in theory provides at least some measure of satisfaction.

Back in the pioneering days of artificial intelligence, in the 1960s, mathematicians and computer scientists such as Claude Shannon used chess as an application to test ways of making computers think more like human beings. Today, complex games of strategy are still used for this purpose. In themselves, of course, the games have no great importance, unless they are how you make a living. But the ways in which machines are constructed or taught, or learn for themselves, to become stronger players can be transferred to other tasks that do matter. What's more, the effort to solve chess and similar complex games sheds some light on the limits of what we can ultimately hope to know.

9

WHAT IS AND WHAT SHOULD NEVER BE

How wonderful that we have met with a paradox. Now we have some hope of making progress.

—NIELS BOHR

Please accept my resignation. I don't want to belong to any club that will accept me as a member.

—GROUCHO MARX

The word "paradox" comes from the Greek *para* ("beyond") and *doxa* ("opinion or belief"). Literally, then, it means anything that's hard to believe or runs counter to our intuition or common sense. We'll often say, in everyday conversation, that something is paradoxical just because it seems almost unbelievable. For example, the fact, mentioned in Chapter 3, that in a room of twenty-three people, there's a fifty-fifty chance of two people having the same birthday is sometimes called the "birthday paradox," even though it's an easily proven statistical fact and surprising only because it jars with our expectations. Among academics in math and logic, the word has a narrower and more precise meaning: it refers to a statement or situation that gives rise to a self-contradiction. One such paradox, as we'll see, led to an important breakthrough in a fundamental area of math. Others, to do with the nature of self, free will, and time, have opened up fruitful discussions in philosophy and science.

Fourteenth-century French priest and philosopher Jean Buridan played an important role in encouraging the Copernican revolution—the idea that the sun is at the center of the solar system—in Europe. But his name is better known through its association with a paradox of medieval logic. Buridan imagined an ass that is standing exactly midway between two piles of hay that are the same in every respect—size, quality, and appearance. The ass is hungry but also unremittingly rational, so it has no reason to favor one pile over the other. Thus conflicted, it simply stays where it is, having no basis on which to make a decision, until it starves. With one pile of food it would have lived, yet with two identical piles it dies. How, if only pure reason is involved, can this make sense?

Buridan's ass is in a predicament similar to that of a perfectly round ball balanced at the top of a steep, rounded hill. As long as no unbalanced forces act on it, there's nothing to cause it to roll down the side. Its state is unstable, in that the slightest nudge will set it in motion. But without a nudge of any kind, it will remain forever in place. Like many thought experiments, Buridan's ass makes a number of assumptions that are never fully realized in practice. For one thing, it assumes complete symmetry: that the decision to choose one pile of hay or the other involves an identical series of states and steps. Yet this would never be the case in reality. The ass might habitually favor its right side over its left or, perhaps through a trick of the light, gain the impression that one pile looks slightly more appetizing than the other. Any of a dozen different reasons might tip the balance in favor of one of the food piles. In a practical example from digital electronics, a logic gate may hang indefinitely midway between the values 0 and 1 (analogous to the bales of hay) until some random flicker of noise in the circuit causes it to flip into one of the stable states. Buridan's ass has been used in discussions of free will, since it's argued that a creature with free will, however rational, would never choose not to eat simply because there was no reason to favor one food source over another.

Another paradox that bears on the problem of free will was devised as recently as 1960 by William Newcomb, a theoretical physicist at the Lawrence Livermore Laboratory and great-grandson of the brother of the famous nineteenth-century astronomer Simon Newcomb. In Newcomb's paradox, a superior being, with predictive powers that have never been known to fail, has put $1,000 in a box labeled A and either nothing or $1 million in a box labeled B. The being presents you with a choice: (1) open box B only, or (2) open both box A and box B. But here's the catch: the being has put money in box B only if it predicted you'll choose option 1. It put nothing in box B if it predicted you'll do anything other than choose that option. The question is, what should you do to maximize your winnings? In fact, there's no known consensus on what to do or even whether the problem is well defined. You might argue that, since your choice now can't alter the contents of the boxes, you may as well open them both and take whatever's there. This seems reasonable until you recall that the being has never been known to predict wrongly. In other words, in some way your mental state is correlated with the contents of the box: your choice is linked to the probability that there's money in box B. These arguments and many others have been put forward in favor of either choice. But there's no generally agreed-upon "right" answer, despite the concerted attentions of philosophers and mathematicians for more than a half century.

Newcomb came up with his paradox while thinking about a slightly older one known as the surprise hanging, which seems to have begun circulating by word of mouth sometime in the 1940s. It concerns a man who's been condemned to hang. A judge, with a reputation for reliability, tells the prisoner on Saturday that he'll be hanged on one of the next seven days but that he won't know (and won't be able to know in any way) which day until he's told on the morning of the execution. Back in his cell, the prisoner thinks about his predicament for a while and then reasons that the judge has made a mistake. The hanging can't be left until the following Saturday because the prisoner would certainly know, if this day dawned, that it was his last. But if Saturday is eliminated, the hanging can't take place on Friday either, because if the prisoner survived Thursday, he would know that the hanging was scheduled for the next day. By the same argument, Thursday can be crossed off, then Wednesday, and so forth, all the way back to Sunday. But with every other day ruled out for a possible surprise hanging, the hangman can't arrive on Sunday without the prisoner knowing in advance. Thus, the condemned man reasons, the sentence can't be carried out as the judge decreed. But then Wednesday morning comes around and, with it, the hangman—unexpectedly! The judge was right after all, and something was awry with the prisoner's seemingly impeccable logic. But what?

More than a half century of attack by legions of logicians and mathematicians has failed to produce a resolution that's universally accepted. The paradox seems to stem from the fact that whereas the judge knows beyond doubt that his words are true (the hanging will occur on a day unknown in advance to the prisoner), the prisoner doesn't have this same degree of certainty. Even if the prisoner is alive on Saturday morning, can he be certain that the hangman will arrive?

Sometimes our use of language, and especially a lack of precision when making statements or asking questions, can lead to perplexing problems. Berry's paradox is named after George Berry, a part-time employee of the Bodleian Library at Oxford University, who, in 1906, drew attention to statements of the form "the smallest number not nameable in under ten words." At first sight, there doesn't seem anything particularly mysterious about this sentence. After all, there are only finitely many sentences that have less than ten words and fewer still that specify unique numbers, so there are clearly a finite number of numbers nameable in under ten words and a smallest number, N, that is not. The trouble is, the Berry sentence itself is a specification for that number in only nine words! In this case, the number N would then be nameable in nine words, contradicting its definition of being the smallest number not nameable in fewer ten words. You could try picking a different number as

N, but the paradox will still hold. What Berry's paradox shows is that the concept of nameability is inherently ambiguous and a dangerous one to be used without qualification.

A paradox of a different kind deals with the notion of identity. We generally take identity for granted; for example, it seems obvious that the person known as Agnijo an hour ago is still the same person now. However, paradoxes can call our intuitive ideas about identity into question. One such paradox involves a thought experiment known as the ship of Theseus. Legendary King Theseus, famous for his association with the story of the Minotaur, fought many successful naval battles so that the people of Athens, it's said, honored him by preserving his ship in port. Over time, however, the planks and other parts of the all-wooden vessel gradually rotted and had to be replaced, one by the one. The question is, at what point, if any, does the ship stop being the ship of Theseus and become instead a replica or a different entity in its own right? After one plank is replaced, or half of all the wood, or some other amount? Does the answer depend on the speed of replacement? If the old planks were then reassembled to form another ship, which one, if any, is the true ship of Theseus?

Such questions may seem unimportant in the case of an inanimate object—although archaeologists and conservationists may debate to what extent ancient buildings and artifacts that have been repaired and rebuilt can be said to be original or legitimate continuations of the original. But thought experiments along the lines of the ship of Theseus take on a new dimension when applied to ourselves and, in particular, to the subject of personal identity. The time is fast approaching when it will be possible to replace almost every body part with an organic transplant (donated or lab grown) or a prosthesis. If a large part of our body is replaced by various means, over time, are we still the same person in the end? The tendency might be to say yes unless the substitution involves significant portions of the brain, because the brain is generally regarded as being key to who we are.

Of course, everyone would agree that if a person loses an arm in an accident and receives a prosthetic arm instead, they remain the same in every way that matters. Also, it's true that the atoms, molecules, and cells that make up our bodies are changing to some extent every moment. In the time it takes you to read this sentence, about 50 million of your cells will have died and been replaced. If it's a like-for-like swap and it happens over time, or we get a transplant or a prosthesis, we don't worry about there being a threat to our identity. We also recognize that people age without becoming someone new. But what if the replacement happened all at once? What if every particle in your body, down to the atomic level, were suddenly swapped out for an identical copy?

Teleportation, in which particles (or, to be more precise, properties of particles) are made to vanish in one place and reappear instantly some distance away, is already possible with photons. It'll probably be a long time before such "quantum teleportation" can be achieved with larger objects. But let's suppose that human teleportation becomes possible. You step onto a teleportation pad in, say, London, the position and state of every atom in your body are scanned in exquisite detail, and a moment later this information is used to reconstitute your body from a new collection of identical atoms in Sydney. The reconstruction is so swift and accurate that, aside from a moment of slight disorientation, you don't notice that your old body in London has been dissolved, the component atoms recycled into the environment, and your new body fashioned a split second later from identical atoms in identical states halfway around the planet. As far as you're concerned, you've just been whisked across more than ten thousand miles in the blink of an eye and can start your Australian adventure without the customary jet lag and tiredness that follow a day's plane journey. You were even thinking exactly the same thought at the instant you were reconstructed as when your old body was dissolved at the London end. Two weeks later, it's time to go home, and you go through the reverse process of having your atoms disassembled in Sydney and an exact copy fashioned a microsecond later in the United Kingdom. You step off the pad, bronzed and relaxed, ready to head home. But at that moment, you get a call on your cell phone from a technician in Australia. There's been a problem at the Sydney end, and the "old" you didn't get dissolved. Instead, he or she is complaining to the staff there that nothing happened, the teleportation failed, and they should either do the whole thing again or offer a refund. So now, it seems, there are two "yous," identical in every respect down to the exact thought and memory at the moment the teleportation took place. Which is the real "you"? And how can you be in two places at once? What happens to your consciousness in such a situation? And what would it *feel* like for a single consciousness to be replicated in this way?

The technological barriers to human teleportation are immense, and it's by no means certain they'll ever be surmounted. However, the feasibility of uploading the contents of our minds into a computer so that we achieve a kind of mental immortality is already being discussed. The ultimate goal would be not only to store all of our memories but also to re-create our consciousness, our active experience of self and the world around us, in an inorganic medium. The issue of what it would mean and feel like to be reconstituted in this way then becomes of central importance. If one copy of your consciousness could be made, then two or more could also be created—backups perhaps, in case the main one was lost or damaged. Such possibilities will raise interesting personal and ethical dilemmas over the coming decades. They'll also forge a direct link between mathematics and the mind. The way the uploading is carried out and the technology of the computational support system required will be the product of intense and complex mathematical analysis, as well as advancements in science and engineering. The outcome, if it happens, will be a new form in which human-level consciousness can exist and be sustained indefinitely. At that point, the ultimate expression of objective universality, drained of emotion and opinion—mathematics—will meet the essence of subjectivity, the feeling of what-it-is-like-to-be.

Time is another great mystery around which the paradoxical swirls. The twins paradox is a thought experiment in which one twin (A) travels into space at nearly the speed of light and returns, after a long interstellar journey, to find that he's aged much less than his twin (B), who remained on earth. The slowing down of time, or time dilation, for an object moving at very high speed is a proven effect of Einstein's special theory of relativity. The riddle posed by the twins paradox is why twin B doesn't also age at a slower rate because he could be considered to be moving equally fast, in the opposite direction, if we switch to a frame of reference in which twin A is at rest. The fact is, though, that the roles played by A and B, despite appearances, aren't symmetric. Twin A had to *accelerate* to reach a high speed, whereas twin B, back on earth, underwent no acceleration. It's this shifting of twin A out of earth's frame of reference that causes him to age at a different rate than his stay-at-home brother.

Traveling very fast is a proven way of jumping into the future, assuming we can develop the technology for super-high-speed travel, though, unfortunately, it's a one-way trip. We don't know of any trick to hop back into the past, except perhaps by means of something exotic (and disturbingly unpredictable), like jumping into a wormhole—a hypothetical tunnel in space and time. But that hasn't stopped people from speculating about what might happen if we *could* make journeys backward through time. One difficulty that might arise is that we end up changing something in the past that renders our future existence problematic. In *Back to the Future,* Marty McFly is hurled back to 1955 in a plutonium-powered DeLorean, only to encounter his mother-to-be as a hormonal teenager and wisely avoids her amorous advances. We might go back and accidentally kill our grandfather when he was still a boy. That would mean we couldn't be born and so couldn't go on to become a time traveler who went into the past and caused the early demise of his grandfather. This "grandfather paradox" is a classic argument against the possibility of going back in time. On the other hand, it's been suggested that if we did go back, we would cause a split in the time line so that whatever we did in the past, as a result of our time-machine exploits, would happen only along a new branch, completely separate from the original, thereby sidestepping any logical conflicts or endless loops.

Such conflicts and loops are not so easily avoided in other cases, however. Suppose these three sentences are written on a card:

1. This sentence contains five words.

2. This sentence contains eight words.

3. Exactly one sentence on this card is true.

Is sentence 3 true or false? Obviously, sentence 1 is true and 2 is false. If 3 is also true, then two sentences are true, which immediately makes 3 false. But if 3 is false, then it isn't true to say that exactly one sentence on the card is true. However, in that case, the only true statement is 1, which means 3 must be true. A statement can't be both true and false at the same time. Can it be neither?

This little conundrum is similar to one credited to sixth-century Greek seer and philosopher-poet Epimenides, who purportedly said, "All Cretans [people from the island of Crete] are liars." Because Epimenides himself was a Cretan, his statement implies that he also is a liar, so that, at first glance, what he says appears to be paradoxical. In fact, though, it isn't, even if we assume that every Cretan either always lies or always tells the truth. Where some people make a mistake is that they know that if Epimenides is truthful, then all Cretans, including himself, are liars (which is a contradiction), but assume that if Epimenides is lying, then all Cretans, including himself, are truthful. This is false, because if Epimenides is lying, this implies only that at least one Cretan is truthful, not necessarily all Cretans.

Epimenides's statement, however, can easily be turned into a genuine paradox. The so-called liar paradox, credited to Eubulides of Miletus, of the fourth century BC, can be put this crisply: "This statement is a lie." It then follows that if it's true, then it's false, and if it's false, then it's true.

Different versions of the basic Eubulides liar paradox have popped up over the centuries. Jean Buridan used it in an argument for the existence of God. Just over a hundred years ago, English mathematician Philip Jourdain offered a version in which two statements are written on opposite sides of a single card. On one side appears this statement: "The sentence on the other side of this card is true." On the other is a baffling counter-statement: "The sentence on the other side of this card is false."

No one has come up with an easy or single resolution of the liar paradox. Common reactions to it are to dismiss it out of hand as being a pointless game of words or to say that the sentence(s) involved, although grammatically correct, are devoid of real content. Both are attempts to stop the paradox dead in its tracks, but don't stand up to scrutiny. The first just refuses to acknowledge that there's any substantive problem. The second denies any meaning to the statement(s) on the grounds that they lead to a paradox. On the face of it, the statement "This statement is a lie" is very similar to the one that declares "This sentence is not in French." How can the first be meaningless if the second makes perfect sense?

Apart from being interesting talking points, such brain twisters don't seem to serve much real purpose. But there's one paradox, leading to a self-contradiction, that has had a pivotal influence on the development of one of the most fundamental areas of modern mathematics. The paradox in question is best understood in a form called the barber paradox. In this, there's a barber who claims to shave everyone who doesn't shave themselves. As a result, he faces a dilemma: Does he shave himself? If he does, he isn't shaved by the barber, so he doesn't shave himself. If he doesn't, he is shaved by the barber, so he *does* shave himself. A more abstract form of this paradox appeared in a letter from English philosopher and logician Bertrand Russell to German philosopher and logician Gottlob Frege in 1902. The timing couldn't have been worse, from Frege's point of view. Frege was just about to send the second volume of his monumental work *Die Grundlagen der Arithmetik* (*The Foundations of Arithmetic*) to the publisher. In his letter, Russell drew attention to a peculiar mathematical object: the set of all sets that don't contain themselves. He then asked, does this set contain itself? If so, then it isn't contained within the set of all sets that don't contain themselves, which means it doesn't contain itself. If not, it *is* contained within the set of all sets that don't contain themselves, which means it does contain itself. Such a monstrosity, Frege realized in horror, could not be accommodated within the set theory he had spent many years formulating and that now, it seemed, was broken and discredited before it had even seen the light of day.

Russell's paradox, as it became known, exposed a fatal inconsistency of the "naive" set theory that Frege had developed. "Naive," in this context, refers to early forms of set theory that aren't based on axioms and that assume there's such a thing as a "universal set"—a set that contains all objects in the mathematical universe. On reading Russell's letter, Frege immediately grasped its implication. In reply to Russell, he said: "Your discovery of the contradiction caused me the greatest surprise and, I would almost say, consternation, since it has shaken the basis on which I intended to build my arithmetic. . . . It is all the more serious since, with the loss of my rule V, not only the foundations of my arithmetic, but also the sole possible foundations of arithmetic seem to vanish." The existence of this one paradox at the heart of Frege's cherished theory meant that, effectively, all the statements the theory could generate were both true and false at the same time. It's a simple fact that any system of logic, if found to harbor a paradox, is rendered useless.

The emergence of Russell's paradox, at the dawn of the twentieth century, shook logic and mathematics to their very core. With a paradox lurking in their midst, no proof they could generate was ultimately trustworthy; no theory rooted in them was well grounded. Operationally, it's true, math could continue as it always had. For everyday purposes, no one was about to deny that 2 + 2 = 4 was true and that 2 + 2 = 5 was just as obviously false. But the disturbing fact remained that there was no way to prove these things, or anything else in math for that matter, starting from what had been assumed to be firm mathematical bedrock—set theory, developed by the likes of Georg Cantor and Richard Dedekind (both of whom we'll learn more about in Chapter 10 on infinity), David Hilbert (whom we first met in Chapter 1, and then again in Chapter 5 in connection with Turing machines), and Frege—in the form in which it existed in late Victorian times. The crumbling of naive set theory began with a paradox to do with transfinite ordinals, known as the Burali-Forti paradox, although it was Cantor who first grasped its disturbing implications in about

1896. Then came Russell with his coup de grâce, and it was clear that mathematicians must either abandon their faith in proof or find an alternative to naive set theory. The former was unthinkable. So some way was needed to rebuild set theory from the ground up but in a way that excluded, from the very start, anything with a whiff of the paradoxical.

The answer lay in the development of what are called formal systems. In contrast to naive set theory, which grew out of commonsense assumptions and rules based on natural language, the new approach started by defining a specific set of axioms. An axiom is a statement or premise that's put forward in precise terms and taken to be true at the outset. Different systems and authors are free to adopt different sets of axioms. But after the axioms have been declared in a formal system, the only statements that can be said to be true or false within the system have to be constructed from those starting assumptions. The key to the success of formal systems is that by carefully choosing axioms in the first place, anything as unwelcome and destructive as the liar paradox can be prevented from arising.

What is sometimes called a paradox may not actually be a paradox but merely a true statement that seems counterintuitive or a false statement that seems obvious. A classic example in mathematics is the so-called Banach-Tarski paradox, which states that you can take a ball, cut it into finitely many pieces, and rearrange them to make two balls, each of the same volume as before. This seems crazy, and, indeed, it's important to understand that this isn't a claim about what can actually be done with a real ball, a sharp knife, and some glue. Nor is there any chance of some entrepreneur being able to slice up a gold ingot and assemble in its place two new ones like the original. The Banach-Tarski paradox tells us nothing new about the physics of the world around us but tells us a great deal about how "volume," "space," and other familiar-sounding things can assume unfamiliar guises in the abstract world of mathematics.

Polish mathematicians Stefan Banach and Alfred Tarski announced their startling conclusion in 1924, having built on earlier work by Italian mathematician Giuseppe Vitali, who proved that it's possible to chop up the unit interval (the line segment from 0 to 1) into countably many pieces, slide these bits around, and fit them together to make an interval of length 2. The Banach-Tarski paradox, which mathematicians often refer to as the Banach-Tarski decomposition because it's really a proof, not a paradox, highlights the fact that among the infinite set of points that make up a mathematical ball, the concepts of volume and measure can't be defined for all possible subsets. What this boils down to is that quantities that can be measured in any familiar sense aren't necessarily preserved when a ball is broken down into subsets and then those subsets reassembled in a different way using just translations (slides) and rotations (turns). These unmeasurable subsets are extremely complex, lacking reasonable boundaries and volume in the ordinary sense, and simply aren't attainable in the real world of matter and energy. In any case, the Banach-Tarski paradox doesn't give a prescription for *how* to produce the subsets; it only proves their existence.

Paradoxes can come in many different forms. Some of them are merely errors in our reasoning; others raise interesting questions about what we may take for granted. Still others can threaten to destroy an entire field of mathematics but provide an opportunity for rebuilding it on more solid foundations.

10

YOU CAN'T GET THERE FROM HERE

> The infinite in mathematics is always unruly unless it is properly treated.
>
> —JAMES NEWMAN

> I cannot help it—in spite of myself, infinity torments me.
>
> —ALFRED DE MUSSET

Does space stop somewhere? Was there a beginning to time, and will it ever end? Is there a biggest number? Even as children we ask these questions. Everyone, it seems, at one time or another, is curious about infinity. Yet infinity, far from being a vague and nebulous concept, can be studied rigorously—and the results we get can be so counterintuitive as to be unbelievable.

Thoughts about endlessness find their way into philosophy, religion, and art. American jazz guitarist and composer Pat Metheny said, "What I look for in musicians is a sense of infinity." English poet and painter William Blake speculated that our senses blocked our appreciation of the true nature of things and that "if the doors of perception were cleansed, everything would appear to man as it is—infinite." French novelist Gustave Flaubert warned of the dangers of thinking too much about the subject: "The more you approach infinity, the deeper you penetrate terror."

Scientists also run into infinity from time to time, and the encounter isn't always a pleasant one. In the 1930s, theorists trying to find better ways to understand subatomic particles found that their calculations led to quantities blowing up—diverging—to infinite values. This happened, for instance, when they treated the electron as a particle having zero size, as electron-electron scattering experiments suggested it did. Their calculations predicted the energy of the electric field around the electron to be infinitely large, which was absurd. Eventually, a way of avoiding this embarrassment was found in the form of a mathematical trick called renormalization, which has now become a standard ploy in the field of quantum mechanics, even though some physicists continue to be uneasy about its arbitrary nature.

At the other end of the physical scale, cosmologists are keen to learn if the universe as a whole is limited in size or is endless in all directions. At the moment, they simply don't know. The part of the universe that we can see (at least in principle)—the so-called *observable* universe—is about 92 billion light-years across, where 1 light-year is the distance that light can travel in a year. The observable universe is the portion of the whole universe from which light has had a chance to reach us since the big bang. Beyond it may lie a much larger, perhaps infinitely large, volume of space to which we have no means of access.

Hubble Space Telescope image of the galaxy cluster Abell S1077. (ESA/HUBBLE, NASA, AND N. ROSE)

Ever since Einstein came up with his general theory of relativity, we've known that the space in which we live can be curved, just as the surface of, for example, a sphere is curved (though our space has three dimensions rather than two). More accurately, spacetime (because space and time are intimately woven together) doesn't have to follow the familiar rules of the geometry we learned in school. On a local scale, we know for sure that spacetime is curved: the spacetime around any object with mass, such as the sun or earth, is warped, just as a sheet of rubber is stretched when a weight is placed on it. But we don't yet know if the universe as a whole is curved (non-Euclidean) or flat. Cosmologists are keen to find out because upon the shape of the universe hinges its ultimate fate.

If, globally, spacetime is curved, then it may be that the universe has a closed shape, in the same way that the surface of a sphere or a doughnut is closed, so that it would be limited in size, yet no matter how far you traveled, you would never reach an edge or boundary. Another possibility is that the universe is shaped something like the surface of a saddle that was continued indefinitely, in which case it could be "open" and stretch away forever or still be finite. It might also, taken as a whole, be flat, in which case it could again be finite or infinite in size. Whatever turns out to be the true situation, if the universe started off finite in size, then it will remain so (although it may keep on growing); if it is infinitely big, then it's always been that way. The idea of a universe that's always been infinite in size seems to be at odds with the popular concept of a big bang that involves the outpouring of matter and energy from a region that started out much smaller than an atom. But there's no inconsistency: this initially tiny region represents only the size of the observable universe—the farthest that light can have traveled—in the fraction of a second since the start of the big bang. The universe as a whole could still be infinite from the outset, though this wouldn't have been observable. Neither option, a universe infinite in space and time or a finite one, sits easily on the mind, but the finite option is perhaps the harder to grasp when we come to think about it. As philosopher and essayist Thomas Paine wrote, "It is difficult beyond description to conceive that space can have no end; but it is more difficult to conceive an end. It is difficult beyond the power of man to conceive an eternal duration of what we call time; but it is more impossible to conceive when there shall be no time."

The evidence that astronomers have gathered so far, from studying distant galaxies, suggests that the universe is both flat and infinite in extent. Just what "infinite" means when it comes to space and time in the real universe, however, isn't obvious. We'll never be able to prove, by direct measurement, that space and time go on forever because we can never receive information from infinitely far away. Another complication arises from the very nature of space and time. Physicists believe that there are such things as a smallest possible distance and time, known as the Planck length and Planck time, respectively. In other words, space and time aren't continuous but instead are granular or quantized. The Planck length is extremely small—a mere 1.6×10^{-35} meter, or 100 million trillionth the width of a proton. And the Planck time, which is how long it takes light to travel a Planck length, is incredibly brief—about 10^{-43} of a second. Nevertheless, the existence of this granularity of spacetime means that we have to be careful when it comes to talking about infinity in the context of the physical universe. Not all infinities are equal, as mathematicians have discovered.

Early Greek and Indian philosophers, more than two thousand years ago, were the first to record their thoughts about infinity. Anaximander, in the sixth century BC, talked about the *apeiron* (boundlessness) as being the source of everything. His countryman Zeno of Elea (now known as Lucania in southern Italy), a century or so later, was the first to deal with infinity from a mathematical perspective.

An early glimpse of the perils of the infinite came to Zeno through his paradoxes, the best known of which pits Achilles in a race against a tortoise. Confident of victory, our mythical hero gives the tortoise a head start. But then, asks Zeno, how can Achilles ever overtake the sluggish reptile? First, he has to catch up to where the tortoise started, by which time the tortoise will have moved on. When he makes up the new distance that separated them, he finds his adversary has advanced again. And so it goes on, indefinitely. No matter how many times Achilles reaches the point where his competitor was, the tortoise has gone a bit farther. Obviously, there's a disconnect here between how we sometimes think about infinity and how things play out in reality. In fact, so perplexed was Zeno by the problem that he decided not only was it best to avoid thinking about the infinite but also that motion was impossible!

A similar shock lay in store for Pythagoras and his followers, who were convinced that everything in the universe could ultimately be understood in terms of whole numbers. After all, even common fractions are just one whole number divided by another. But the square root of 2—the length of the diagonal of a right-angled triangle whose shorter sides are both one unit long—refused to fit into this neat cosmic scheme. It was an "irrational" number, inexpressible as the ratio of two integers. Put another way, its decimal expansion goes on forever, without ever settling into a repeating pattern. The Pythagoreans knew nothing of this, only that the square root of 2 was a monstrosity in their seemingly perfect worldview, and so they tried to keep its existence a secret.

These two examples highlight a basic problem in coming to grips with infinity. Our imaginations can cope with something that hasn't yet reached an end: we can always picture taking another step or adding one more to a total. But infinity, taken as a whole, fully formed, boggles the mind. For mathematicians, this was a major headache because their subject deals with precise quantities and meticulously well-defined concepts. How could they work with things that clearly existed and went on indefinitely—a number like $\sqrt{2}$ (which starts 1.41421356237 . . . and just goes on and on without any predictable pattern or repetition) or a curve that approached a line ever more closely—while avoiding a confrontation with infinity itself? Aristotle offered a possible solution by arguing that there were two kinds of infinity. *Actual infinity*, or *completed infinity*, which he believed couldn't exist, is endlessness fully realized, infinity actually attained (mathematically or physically), at some point in time. *Potential infinity*, which Aristotle insisted was evident in nature—for example, in the unending cycle of the seasons or the indefinite divisibility of a piece of gold (he didn't know about atoms)—is infinitude spread over unlimited time. This fundamental distinction, between absolute and potential infinities, lasted in mathematics for more than two thousand years.

In 1831 no less a figure than Carl Gauss expressed his "horror of the actual infinitude," saying, "I protest against the use of infinite magnitude as something completed, which is never permissible in mathematics. Infinity is merely a way of speaking, the true meaning being a limit which certain ratios approach indefinitely close, while others are permitted to increase without restriction."

By confining their attention to potential infinity, mathematicians were able to deal with and develop crucial concepts, such

Looking to infinity. A selfie of a selfie of a selfie . . . by David using a mirror and cell phone. (DAVID DARLING)

as those of infinite series, limits, and infinitesimals, and so arrive at the calculus, without having to grant that infinity itself was a mathematical object. Yet as early as the Middle Ages, certain paradoxes and puzzles arose, which suggested that actual infinity wasn't an issue to be easily brushed aside. These puzzles stem from the principle that it's possible to pair off all the members of one collection of objects with all those of another of equal size. Applied to indefinitely large collections, however, this principle seemed to fly in the face of a commonsense idea first expressed by Euclid, namely, that the whole is always greater than any of its parts. For instance, it appeared possible to pair off *all* the positive integers with *only those that are even*: 1 with 2, 2 with 4, 3 with 6, and so on, despite the fact that positive integers also include odd numbers. Galileo, in considering such a problem, was the first to show a more enlightened attitude toward the infinite when he said, "Infinity should obey a different arithmetic than finite numbers."

The concept of potential infinity lulls us into thinking that if we just keep going far enough, or long enough, we'll get closer to infinity. It's then only a small step to the popular myth that infinity is just like a very big number and that a trillion, or a trillion trillion trillion, is somehow closer to infinity than, say, ten or a thousand. But this simply isn't the case. Traveling farther down the number line or counting to bigger and bigger numbers gets us *no nearer whatsoever* to infinity. We are just as far from, or close to, infinity at the number 1 as we are at any finite number we care to name, however vast. Put another way, all of infinity is contained within every single number, no matter how small, so that setting off down the road to larger and larger numbers in search of infinity is utterly futile. The fact is that infinity exists between 0 and 1, for instance, because there are infinitely many fractions—½, ⅓, ¼, and so on. Infinity is not like a big finite number at all. To deal with infinity, we have to jump out of the realm of finite numbers altogether and stop using them as a crutch to our understanding.

German mathematician David Hilbert offered a striking illustration of how weird the arithmetic of the endless can get.

Imagine, said Hilbert in a 1924 lecture, a hotel with an infinite number of rooms. In the usual kind of hotel, with finite accommodation, no more guests can be squeezed in once all the rooms are full. But "Hilbert's Grand Hotel" is dramatically different. If the guest occupying room 1 moves to room 2, the occupant of room 2 moves to room 3, and so on, all the way down the line, a newcomer can be placed in room 1. In fact, space can be made for an infinite number of new clients by moving the occupants of rooms 1, 2, 3, and so on, to rooms 2, 4, 6, and so on, thus freeing up all the odd-numbered rooms. This process can be continued indefinitely so that even if an infinite number of coaches were to arrive, each carrying an infinite number of passengers, no one would have to be turned away. It's a result that makes a mockery of our intuition, but our intuition isn't used to dealing with things that are infinitely large. The fact is that the properties of infinitely many things are different from those of finitely many things, just as, for example, in science the world behaves differently at a very small (quantum) scale than it does at an everyday level. In the case of Hilbert's Grand Hotel, the statements "There is a guest to every room" and "More guests can be accommodated" aren't mutually exclusive.

Such is the bizarre world we enter if we accept the reality of sets of numbers with infinitely many elements. That was a crucial issue facing mathematicians in the late nineteenth century: Were they prepared to embrace actual infinity as a number? Most still stood behind Aristotle and Gauss in opposing the idea. But a few, including German mathematician Richard Dedekind and, above all, his compatriot Georg Cantor, understood that the time had come to put the concept of infinite sets on a sound logical footing.

In pioneering the strange and unsettling realm of the infinite, Cantor, faced with fierce opposition and derision from many of his contemporaries (including, most damningly, his old mentor and teacher Leopold Kronecker), lost his job at the University of Berlin and, from time to time, his sanity. In later life, he was an occasional patient in mental institutions, agonized over the authenticity of Shakespeare's writings, and became embroiled with the philosophical and even religious implications of his mathematical work. Although he died miserably in a sanatorium in 1918, with his nation still at war, he's now remembered for his fundamental contributions to set theory and our understanding of the infinite.

Cantor realized that the well-known pairing-off principle used to tell if two finite sets are equal could be applied equally well to infinite sets. It followed that there really are just as many even positive integers as there are positive integers altogether. Far from being a paradox, he saw, this was a *defining property* of infinite sets: that the whole is no bigger than some of its parts. He went on to show that the set of all natural numbers, which is the set of all nonnegative integers, 0, 1, 2, 3, . . . (sometimes 0 is not included), contains precisely as many members as the set of all rational numbers—numbers that can be written as one whole number divided by another. He called this infinite number aleph-null ($\aleph_0$), "aleph" being the first letter of the Hebrew alphabet and "null" the German for "zero." Sometimes it's also called aleph-zero or aleph-nought.

You might suppose there could be only one infinite number, because once something is endlessly big, how could anything be bigger? But you would be wrong. Cantor showed that there are different kinds of infinity of which aleph-null is the smallest. Infinitely bigger than aleph-null is aleph-one (which he described as having greater "mightiness"), infinitely bigger again than aleph-one is aleph-two, and so on, without end. Unhelpfully as far as the imagination goes, alephs come in infinitely many sizes. Not only that, but corresponding to each aleph, it turns out, are infinitely many other infinite numbers—a fact that leads us to consider the important difference, in the realm of the infinite, between cardinals and ordinals.

In everyday language and arithmetic, cardinal numbers tell us how many there are in a collection of things (1, 5, 42), whereas ordinal numbers, as the name suggests, tell the order or position of something (1st, 5th, 42nd). The distinction between these two types of numbers seems pretty clear and not that important. Say we're talking about pencils. It's obvious that you can't have a fifth pencil without having at least five pencils in a group and that you could still have a fifth pencil even if there were seven in a group. You could also have five pencils without having a fifth pencil, if you didn't put them in any order. But these little distinctions aside, we can use the same symbols for cardinals and ordinals—1 (or 1st), 5 (or 5th), 42 (or 42nd), and so on—and not be too fussed over how cardinals and ordinals differ. Cantor realized that when it comes to infinite numbers, however, the distinction becomes vitally important. To appreciate this, we need to take a quick look at an area of math that Cantor and Dedekind were instrumental in developing, namely, set theory.

A set is just a collection of things, which might be numbers or anything else, and the mathematical symbol used to show a set is a pair of braces or curly brackets. For instance,{1, 4, 9, 25} and {arrow, bow, 75, R} are both sets. The size of a set—how many elements it contains—is known as its cardinality and is given by a cardinal number. Both the sets just mentioned have four members, or elements, and so have a cardinality of four. In general, if the cardinality of two sets is the same, then every member in the first set can be paired off with one in the second so that nothing is left over; in other words, they have a one-to-one correspondence. For example, we can pair 1 with 75, 4 with arrow, 9 with R, and 25 with bow to show that these sets have the same cardinality. The finite cardinals—the cardinals that measure the size of finite sets—are just the natural numbers 0, 1, 2, 3, and so on. The first infinite cardinal is aleph-null, which, as we saw earlier, measures the size of the set of all natural numbers.

For finite sets, the difference between the size of a set, given by a cardinal number, and its "length," given by an ordinal number, is so slight as to be almost pedantic. But in the case of infinite sets, Cantor realized, these are two very different animals. To grasp how different they are, we need to understand the idea of a "well-ordered" set. A set is considered to be well ordered if it satisfies two conditions: first, it must have a definite first member, and, second, each subset, or subgrouping, of its members must also have a first member. The finite set {0, 1, 2, 3}, for instance, is well ordered. The set of all integers, on the other hand, which includes all negative whole numbers as well as all positive ones, {0 . . . –2, –1, 0, 1, 2, . . .}, isn't well ordered because there's no first member. The set of all natural numbers, {0, 1, 2, 3, . . .}, is well ordered because despite having no specified member at the end, it has one at the start, and every subset containing only natural numbers also has a first member.

Now, a key point is that well-ordered infinite sets of the same size, or cardinality, *can have different lengths*. That isn't an easy concept to grasp, even for a mathematician. Strictly speaking, we should say different "ordinalities" rather than "lengths," but the more familiar term helps to appreciate what's going on. Think about the sets {0, 1, 2, 3, 4, . . .} and {0, 1, 2, 4, . . . , 3}, where the three dots mean "carry on forever," starting at 4 and going onward but, in the second set, leaving 3 to the very end. Both sets contain all the natural numbers and therefore have the same size or cardinality, aleph-null. But the second is slightly longer. At first, this doesn't seem to make sense. After all, if we were talking about finite sets, then it's obvious that {0, 1, 2, 3, 4} and {0, 1, 2, 4, 3} are identical in length because they both contain five members. But infinite sets are fiendishly counterintuitive. The set {0, 1, 2, 3, 4, . . .} has no finite end member because the three dots tell you to carry on forever without stopping. However, {0, 1, 2, 4, . . . , 3} is different. It too contains a sequence of members that carries on forever. However, it also contains one member that is beyond all the members of the never-ending sequence. With the 3 taken out, the sequence 0, 1, 2, 3, . . . is just as long as 0, 1, 2, 4, . . . ; in other words, you could pair off all the members of these two sequences and never have one left over. But moving the 3 to the end, so that it comes after the infinite sequence, adds one to the length. Think of it another way. With the first set, {0, 1, 2, 3, 4, . . .}, there's a first element (0), a second element (1), a third element (2), a fourth element (3), and so on. With the second, there's also a first (0), second (1), third (2), fourth (4), and so forth. However, there's one element, 3, that is none of these. The ordinal we assign to 3—not the value of the number but the order in which it appears—is greater than anything that comes before it, because it comes after every other number in the sequence.

We need a naming system for this class of infinite numbers, which is different from the alephs. Mathematicians call the smallest infinite ordinal—the shortest length of the set of all natural numbers—"omega" (ω). The ordinality of the set {0, 1, 2, 4, . . . , 3}, where the 3 is placed after all the other natural numbers, is one greater, namely, $\omega + 1$. Another way of saying this is that 3 is the ($\omega + 1$)th element in the set {0, 1, 2, 4, . . . , 3}. The + sign here is a bit confusing because it doesn't mean addition in the usual sense but, rather, that $\omega + 1$ is the next ordinal after ω. We can add to ω, but we can't take away. The ordinality of {0, 1, 2, 4, . . .}, with the 3 taken out, is still ω. There's no such thing as $\omega - 1$, which may seem strange, but that's because we're used to dealing with finite numbers. The fact is that the "length" of the set of all natural numbers can't be reduced, no matter how large a finite number of elements is removed from it, by virtue of its infinity, as you can see with {0, 1, 2, 4, . . .}. On the other hand, its length *can* be increased by putting the elements that have been removed at the end.

To recap: Aleph-null and ω both refer to the same set—the set of natural numbers. Aleph-null is its size (how many elements it contains), and ω is its shortest length. This length can be increased by taking elements out of their usual order and placing them at the end. The set {2, 3, 4, . . . , 0, 1}, for instance, has a cardinality of aleph-null but an ordinality of $\omega + 2$. We can keep on increasing the length of the set of natural numbers by moving more and more elements beyond the three dots that mean "carry on forever": $\omega + 3$, $\omega + 4$, . . . , all the way up to

$\omega + \omega$ (or $\omega \times 2$), which could be written, for instance, as the subset of all even numbers followed by the subset of all odd numbers, {0, 2, 4, . . . , 1, 3, 5, . . .}, since each of these is equal in length to ω. Then we can continue as before by shifting elements to the end; for instance, one way to write $\omega \times 2 + 1$ is {2, 4, . . . , 1, 3, 5, . . . , 0}. Then we can move on to powers of ω, such as ω^2, ω^3, . . . , all the way up to ω to the power of ω (ω^ω), and then to stacks of powers (power towers) of ω, stretching higher and higher until we reach a power tower of ωs that is ω high. Finally, beyond this lies a new level—an ordinal that Cantor called epsilon-zero (ε_0). Just as ω is the smallest ordinal that lies beyond the finite ordinals, ε_0 is the smallest ordinal that lies beyond any ordinal that can be expressed in terms of ω using addition, multiplication, and exponentiation. It's the gateway to the realm of epsilon numbers, which, like that of the omega ordinals, is infinitely large. The whole process described for the omegas repeats for the epsilons until all the mathematical operations that are possible using epsilons, including power towers of epsilons or even epsilons of epsilons, are exhausted. At this point, we arrive at yet another level of infinite ordinals, starting with zeta-zero (ζ_0). And so it goes, on and on and on.

More than anything, the difficulty in progressing further is one of notation. Eventually, all the Greek letters are exhausted, along with any other ordinary labeling system, to represent the hierarchy of infinite ordinals that stretches away into the distance. Compounded with the problem of finding more powerful and compact means of notating vast infinite ordinals is a mounting degree of technical difficulty. Some milestones, named after the mathematicians with whose work they're associated, lie along the way, once zeta-zero has been left far behind: the Feferman-Schütte ordinal, the small and large Veblen ordinals (both of which are outrageously large), the Bachmann-Howard ordinal, and the Church-Kleene ordinal (first described by American mathematician Alonzo Church and his student Stephen Kleene). To describe properly what any of these mean would take a book in itself, so esoteric is the math behind them. The Church-Kleene ordinal, for instance, is so incomprehensibly vast that there's *no* notation whatsoever that can reach up to it.

These ordinals are rarely encountered by professional mathematicians, let alone the wider public, but the essential point about them all is that they're *countable*. In other words, all the infinite ordinals we've talked about so far, starting with ω, can be paired off one to one, leaving none left over, with the natural numbers, which makes sense, as all of those sequences are merely rearrangements of natural numbers. Another way of saying this is that they all correspond to the cardinality—the size—of aleph-null. We're no nearer to a bigger kind of infinity when we get to epsilon-zero or even the mighty Church-Kleene ordinal than when we started: colossal though they may be, they merely represent different ways of ordering the set of all natural numbers. A bigger kind of infinity means one that transcends aleph-null altogether. But how is that possible?

Aleph-null doesn't behave like the numbers we're used to dealing with. Whereas 1 + 1 = 2, aleph-null + 1 is still aleph-null. Aleph-null plus any finite number or minus any finite number is still aleph-null. This suggests a new twist to the song "Ten Green Bottles," along the lines: "Aleph-null green bottles hanging on the wall, aleph-null green bottles hanging on the wall, and if one green bottle should accidentally fall, there would be aleph-null green bottles hanging on the wall" (repeat ad infinitum). You can't change aleph-null by subtracting from it, adding to it, or multiplying it by any finite number or even multiplying it by aleph-null itself. But Cantor showed, using a theorem that's now named after him, that there's a hierarchy of infinities of which aleph-null is the smallest. The next infinite cardinal, aleph-one, is much bigger and equal in size to the set of all countable ordinals, namely, those with cardinality aleph-null. It's hard to explicitly show ordinals with size aleph-one as a sequence, although the sequence $\{0, 1, 2, \ldots, \omega, \omega + 1, \ldots, \omega \times 2, \ldots, \omega^2, \ldots, \omega^\omega, \ldots, \varepsilon_0, \ldots\}$, and so on, listing every countable ordinal (every different possible length that can be obtained by rearranging the natural numbers), would have ordinality omega-one (the smallest ordinal corresponding to aleph-one).

A quick reminder of what "countable" means: simply a sequence or a set that can be counted. What this means is that it can be put into a sequence, not necessarily in its normal order. Sometimes some shuffling is required, as in Hilbert's Hotel. Because the natural numbers are countable, aleph, the size of the set of natural numbers, is said to be a countably infinite cardinal. Corresponding with it is the smallest infinitely countable ordinal, ω, and infinitely many other countably infinite ordinals. All of these infinitely many countable ordinals arise because, in the case of ordinals, information on order matters so that a much finer distinction has to be made than with cardinals. Even so, all the countable ordinals from ω onward, including the epsilon numbers and the rest, fall within the same cardinality—aleph-null. But with aleph-one comes a dramatic change. Not only is aleph-one indescribably larger than aleph-null, but it's also *uncountable*. Corresponding to it is the smallest uncountable ordinal: omega-one (ω_1).

We've said that aleph-one is the size of the set of countable ordinals, but does it have any other interpretation? Aleph-null measures the size of the set of all natural numbers. Does aleph-one also correspond with anything that's familiar and conceptually easy to grasp? Cantor thought so. He believed that aleph-one was identical with the total number of points on a mathematical line, which, astonishingly, he found was the same as the number of points on a plane or in any higher-dimensional space. This infinity of spatial points, known as the *power of the continuum*, *c*, is also the set of all real numbers (all rational numbers plus all irrational numbers). The real numbers, unlike the natural numbers, can't be counted. Say I were to ask you what comes next after 357 in the real numbers sequence. You could rearrange the real numbers and form as many counting strategies as you want, but the fact remains that there would be real numbers that you could never count, even if you kept counting forever.

Cantor put forward an important idea that came to be known as the continuum hypothesis (CH). According to this, *c* equals aleph-one, which is equivalent to saying that there's no infinite set with a cardinality between that of the natural numbers and that of the real numbers. Yet despite much effort, Cantor was never able to prove or disprove his hypothesis. We now know why—and it strikes to the very foundations of mathematics.

In the 1930s, Austrian-born logician Kurt Gödel showed that it's impossible to prove the continuum hypothesis is wrong starting out from the standard axioms, or assumptions, of set theory.

In order to do this, Gödel put together an explicit system of sets, called the constructible universe, in which he showed that all the axioms hold and the continuum hypothesis is true (although it doesn't follow that the constructible universe is the *only* such system). Three decades later, American mathematician Paul Cohen showed that neither can it be proved correct from those same axioms. Its status was, in other words, indeterminate within the normal framework that mathematicians used. Such a situation had been in the cards ever since the emergence of Gödel's famous incompleteness theorem, which we first met in Chapter 5 and which states that in every sufficiently complex system of axioms, if the system is complete, then there exist statements that can be neither proven nor disproven (more on this when we come back to the incompleteness theorem in Chapter 13). But the independence of the continuum hypothesis was still unsettling, because it was the first concrete example of an important question that provably couldn't be decided either way from the universally accepted system of axioms on which most of mathematics is built.

The debate about whether the continuum hypothesis is ultimately true, or whether it's even a meaningful statement, rumbles on among mathematicians and philosophers. As for the nature of the various types of infinities and the very existence of infinite sets, they depend crucially on what number theory is being used. Different axioms and rules lead to different answers to the question "What lies beyond all the integers?" This can make it difficult or even meaningless to compare the various types of infinities that arise and to determine their relative size, although within any given number system, the infinities can usually be put into a clear order.

There is a towering hierarchy of cardinals beyond aleph-null. Assuming the continuum hypothesis to be true, which is the default position of most mathematicians (because it has helpful consequences), the next biggest infinite cardinal is aleph-one, equal to the size of the set of all real numbers, or, alternatively, all the different ways of ordering the members of aleph-null. After this comes aleph-two, which is equal to how many different ways you can order the members of aleph-one, then aleph-three, aleph-four, and so on, without end. To each aleph corresponds an infinite number of ordinals, the smallest of which is ω in the case of aleph-null, ω_1 in the case of aleph-one, ω_2 in the case of aleph-two, and so on.

Even though there are infinitely many alephs, each infinitely bigger than the one before, mathematicians can dream of cardinals whose size exceeds that of any conceivable aleph. To do this, they have to move beyond the usual foundations of their subject and resort to what are called forcing axioms—a technique pioneered by Paul Cohen, mentioned earlier. This leads to the concept of the modestly named "large cardinals," which in reality are spectacularly vast, including those with special names such as Mahlo cardinals and supercompact cardinals.

Finally (at least for now), there's the notion of absolute infinity, sometimes represented by capital Omega (Ω)—an infinity that transcends or surpasses all others. Cantor himself spoke about it but mainly in religious terms. He was a deeply committed Lutheran whose Christian convictions occasionally surfaced in his scholarly work. To him, Omega, if it existed, could do so only in the mind of the God in which he believed. On that basis, Omega is nothing more than a grand metaphysical speculation.

Sticking purely to mathematics, absolute infinity can't be defined rigorously, so mathematicians, unless they allow philosophical speculation to get the better of them, tend to ignore it. There may be the temptation to characterize it as the number of elements in the universe of all sets—the so-called von Neumann universe. But the von Neumann universe isn't actually a set (rather, it's a class of sets), so it can't be used to define any specific kind of infinity, whether cardinal or ordinal. More controversially, Omega might be thought of as the most sensible result if 1 were to be divided by 0. This isn't a procedure normally defined in math, though it can be done in certain forms of geometry, such as projective geometry, which yield the idea of a "point at infinity" or a "line at infinity." The quest for Omega will continue to challenge future generations of mathematicians, logicians, and philosophers. Meanwhile, we have plenty of infinities, each infinitely larger than the one before, to keep our brains occupied.

One final thought: Are any of these mathematical infinities enacted in the real world, or are they pure abstractions? We saw earlier that cosmologists are leaning toward the view that the universe we live in is geometrically flat and endless in space and time. If it does go on forever, with which kind of mathematical infinity does it correspond? The fact that space and time appear to come in discrete amounts—the Planck length and Planck time—means that they're not continuous like the points on a mathematical line. So if the actual universe is infinitely large, it seems that it could correspond with only the smallest kind of infinity, aleph-null. Anything bigger may always be confined to our intellects or some Platonic space unfettered by the laws of physics.

11

THE BIGGEST NUMBER OF ALL

> The trouble with integers is that we have examined only the very small ones. Maybe all the exciting stuff happens at really big numbers, ones we can't even begin to think about in any very definite way.
>
> —RONALD L. GRAHAM

Ask a child what's the biggest number they can think of, and the answer often runs along the lines of "50 thousand million billion trillion trillion . . . " until they run out of breath, with the odd, nebulous "kazillion" or "bazillion" thrown in for good measure. Such numbers can certainly be big by everyday standards—maybe more than all the living things on earth or all the stars in the universe. But they're peanuts compared with the kinds of mind-bogglingly huge numbers that mathematicians can come up with. Even if you were determined and foolish enough to spend your entire adult waking life saying "trillion" every second, the number you would have named in the end would be unbelievably tiny compared to the monsters of the numerical cosmos we're about to meet, such as Graham's number, TREE(3), and the seriously colossal Rayo's number.

One of the first people known to have thought seriously about very large numbers was Archimedes, born in Syracuse, Sicily, around 287 BC and widely considered to be the greatest mathematician of ancient times and one of the greatest in history. He wondered how many grains of sand there were in the world and, beyond that, how many could be crammed into the whole of space, which the ancient Greeks believed stretched out as far as a sphere holding what they called the fixed stars (in other words, the stars in the night sky as distinct from the planets). His treatise *The Sand-Reckoner* begins: "There are some, King Gelon, who think that the number of the sand is infinite in multitude; and I mean by the sand not only that which exists about Syracuse and the rest of Sicily but also that which is found in every region whether inhabited or uninhabited. Again there are some who, without regarding it as infinite, yet think that no number has been named which is great enough to exceed its multitude."

To pave the way for his cosmos-wide sand estimate, Archimedes set about extending the system available at the time for naming big numbers—the key challenge that has faced all mathematicians ever since who have tried to define larger and larger integers. The Greeks referred to 10,000 as *murious*, which translates as "uncountable" and which the Romans called *myriad*. As the starting point for his venture into the realm of truly huge numbers, Archimedes used a "myriad

myriad," that is, 100 million, or, in modern exponential notation, 10^8—a number far bigger than anything for which the Greeks had a practical purpose. Any number up to a myriad myriad Archimedes labeled as being of the "first order." Numbers up to a myriad myriad times a myriad myriad (1 followed by 16 zeros, or 10^{16}) he said were of the "second order," and so he went on in this way to the third order and the fourth, each order in his scheme being a myriad myriad times larger than the numbers of the previous order. Eventually, he reached numbers of the myriad myriadth order, in other words 10^8 multiplied by itself 10^8 times, or 10^8 raised to the power 10^8. By this process, Archimedes was able to describe numbers up to those with 800 million digits. All these numbers he defined as belonging to the "first period." The number $10^{800,000,000}$ itself, he took to be the starting point for the second period, and then he began the process all over again. He defined orders of the second period by the same method as before, each new order being a myriad myriad times greater than the last, until, at the end of the myriad myriadth period, he had reached the colossal value of $10^{80,000,000,000,000,000}$, or a myriad myriad raised to the power of a myriad myriad times a myriad myriad.

As it turned out, for his sand-counting endeavor, Archimedes needn't have bothered to go beyond the first of his periods. In his cosmic scheme of things, the whole of space, out as far as the fixed stars, was the equivalent of two light-years in diameter, with the sun at its center. Using his estimate of the size of a grain of sand, he came up with a figure of 8×10^{63} grains needed to turn the cosmos into one giant beach—a number of only the eighth order of the first period. Even taking a modern estimate for the diameter of the observable universe of

How many grains of sand on a beach? In the whole world? (DAVID DARLING)

92 billion light-years, there would be no room for more than about 10^{95} grains of sand, which is still a number of merely the twelfth order, first period.

Archimedes may have been the Wizard of the West as far as large numbers went, but in the East intellectuals would soon take the quest for numerical behemoths much further. In an Indian text called the *Lalitavistara Sutra*, written in Sanskrit and dating from about the third century, Gautama Buddha is portrayed as describing to mathematician Arjuna a system of numerals beginning with the *koti*, a Sanskrit term for 10 million. From this starting point, he works his way through a long list of named numbers, each 100 times bigger than the last:

100 *koti* make an *ayuta*, 100 *ayuta* make a *niyuta*, and so on, all the way up to the *tallakshana*, which is one followed by fifty-three zeros. He also gave names to even bigger numbers, such as the *dhvajhagravati*, equal to 10^{99}, all the way up to the *uttaraparamanurajapravesa*, which is 10^{421}.

Another Buddhist text went further—spectacularly further—along the road to the eye-wateringly vast. The *Avatamsaka Sutra* describes a cosmos of infinitely many interpenetrating levels. In Chapter 30, the Buddha is once again expounding on big numbers, starting from 10^{10}, squaring this to give 10^{20}, squaring this in turn to give 10^{40}, and so on to 10^{80}, 10^{160}, 10^{320}, until he reaches $10^{101,493,392,610,318,652,755,325,638,410,240}$. Square this, he declared, and the number reached is "incalculable." Beyond that, having apparently raided the Sanskrit thesaurus for superlatives, he names successively larger numbers "measureless," "boundless," "incomparable," "innumerable," "unaccountable," "unthinkable," "immeasurable," and "unspeakable," before culminating with "untold," which works out to be $10^{10\times(2\wedge122)}$. This dwarfs the biggest number that Archimedes entertained in his writings—$10^{80,000,000,000,000,000}$—to the extent that Archimedes's number would have to be raised to the power of roughly 66,000,000,000,000,000 to get into even the same ballpark as "untold."

Both Archimedes and the Buddhist sutras used large numbers to give some impression of the vastness of their respective versions of the universe. In the Buddhist case, too, it was believed that naming a thing imbued one with a certain power over it. But mathematicians generally weren't too interested in inventing schemes for naming and representing bigger and bigger numbers just for the sake of it. Our convention of using words ending in "-illion" to name big numbers dates back to fifteenth-century French mathematician Nicolas Chuquet. In an article, he wrote down a huge number, broke it down into groups of six digits, and then suggested that the groups be called "million, the second mark byllion, the third mark tryllion, the fourth quadrillion, the fifth quyillion, the sixth sixlion, the seventh septyllion, the eighth ottyllion, the ninth nonyllion and so on with others as far as you wish to go."

In 1920 American mathematician Edward Kasner asked his nine-year-old nephew, Milton Sirotta, to invent a name for the number 1 followed by 100 zeros. His suggestion, "googol," entered the popular lexicon after Kasner wrote about it in the book *Mathematics and the Imagination*, coauthored with James Newman. The young Sirotta also offered the name "googolplex" for the number "one, followed by writing zeroes until you get tired." Kasner opted for a more precise definition "because different people get tired at different times and it would never do to have Carnera [a heavyweight boxing champion] be a better mathematician than Dr Einstein, simply because he had more endurance." However, the actual effect of writing a googolplex is accurate, if a massive understatement. A googolplex, as Kasner defined it, is 10^{googol}, or 1 followed by a googol number of zeros. Whereas a googol is easy to write out in full:

10,000,000,000,000,000,000,000,000,000,000,000,000,000,
000,000,000,000,000,000,000,000,000,000,000,000,000,
000,000,000,000,000,000,000,000,000

a googolplex is sensationally larger. There isn't enough paper on earth, or, if it comes to that, matter in the entire observable universe, to write out the digits of a googolplex, not even if you wrote the zeros as small as protons or electrons. The googolplex is larger than any named number of antiquity, including the mighty "untold." However, it is smaller than a number that, in 1933, arose out of some research being carried out by South African mathematician Stanley Skewes into prime numbers. What became known as Skewes' number is an upper bound, or maximum possible value, to a problem about how prime numbers are distributed. G. H. Hardy, famous British mathematician, mentor of Ramanujan, and author of the widely read *A Mathematician's Apology*, described Skewes' number at the time as "the largest number which has ever served any definite purpose in mathematics." It has the value $10^{10\wedge 10\wedge 34}$ or, more precisely, $10^{10\wedge 8{,}852{,}142{,}197{,}543{,}270{,}606{,}106{,}100{,}452{,}735{,}038.55}$. This colossal upper bound itself required the assumption of the Riemann hypothesis, which, as we saw in Chapter 7, still continues to stump mathematicians. A couple of decades later, Skewes announced another number, in connection with the same prime-numbers problem, but without assuming the Riemann hypothesis, that was even bigger—$10^{10\wedge 10\wedge 964}$, give or take a few trillion.

Not to be outdone by pure mathematics, physics was also coming up with its own gigantic numbers as solutions to some unusual conundrums. An early player in the big-numbers game on the physics front was French mathematician, theoretical physicist, and polymath Henri Poincaré, who, among his many contributions, wrote about how long it would take physical systems to return, exactly, to a certain state. In the case of the universe, the so-called Poincaré recurrence time is the interval it would take for matter and energy to rearrange itself, down to the subatomic level, having gone through an unbelievable number of combinations in between, back to a certain starting point. Canadian theorist Don Page, onetime student of Stephen Hawking, has estimated that the Poincaré recurrence time for the observable universe is $10^{10\wedge10\wedge10\wedge2.08}$ years, which is a number lying somewhere between the small and large Skewes' numbers and larger than the googolplex. He also calculated the maximum Poincaré recurrence time for any universe of a particular type, which was even larger, being $10^{10\wedge10\wedge10\wedge10\wedge1.1}$ years—larger than the large Skewes' number. As for the googolplex itself, Page has pointed out that it's roughly equal to the number of microscopic states in a black hole with a mass as big as that of the Andromeda galaxy.

Untold, the googolplex, and the Skewes' numbers are all colossal in terms of what we can truly grasp in our minds. But they're vanishingly small compared with a number named after American mathematician Ronald Graham, who first published it in a paper in 1977. Like the Skewes' numbers before it, Graham's number arose in connection with a serious mathematical problem, in this case to do with a branch of the subject called Ramsey theory. Approaching Graham's number has to be done in stages, just as if you were climbing one of the world's great mountains. The first step is to be aware of a way of representing very large numbers, devised by American computer scientist Donald Knuth, which is known as up-arrow notation. This builds on the idea that multiplication can be thought of as repeated addition, and exponentiation (raising a number to a power) can be thought of as repeated multiplication. For example, 3×4 is $3 + 3 + 3 + 3$ and $3^4 = 3 \times 3 \times 3 \times 3$. In Knuth's notation, exponentiation is shown by a single up arrow: for example, a googol, or 10^{100}, is written as $10\uparrow100$, and 3 cubed, or 3^3, becomes $3\uparrow3$. Repeated exponentiation, for which we have no everyday notation, is shown by two up arrows, so that $3\uparrow\uparrow3 = 3^{3\wedge3}$. The $\uparrow\uparrow$ operation, known as tetration (because it comes fourth in the hierarchy after addition, multiplication, and exponentiation), is a lot more powerful than it seems at first glance. $3\uparrow\uparrow3 = 3^{3\wedge3} = 3^{27}$, which has the value 7,625,597,484,987.

Another way to show tetration is in the form of a power tower, which is a typesetter's worst nightmare. If a number *a* is to be tetrated to order *k*, then it can be written:

$$a \uparrow\uparrow k = \underbrace{a \uparrow (a \uparrow (\ldots \uparrow a))}_{k \text{ copies of } a} = \underbrace{a^{a^{\cdot^{\cdot^{\cdot^{a}}}}}}_{k \text{ copies of } a}$$

In other words, the number *a* raised to a stack of exponents $k-1$ high.

The rate at which the operators gather momentum is startling: $3 \times 3 = 9$, $3\uparrow3 = 27$, $3\uparrow\uparrow3$ is more than 7.6 trillion (a thirteen-digit number). Tetrating 4 is even more surprising: $4\uparrow\uparrow4 = 4\uparrow4\uparrow4\uparrow4 = 4\uparrow4\uparrow256$, which is roughly $10\uparrow10\uparrow154$—a number bigger than a googolplex ($10\uparrow10\uparrow100$). We've gone beyond the mighty googolplex with nothing more than a simply written action on the number 4.

The giant step from exponentiation to tetration suggests that adding another up arrow will yield something even more dramatic, and we're not disappointed. Repeated tetration, called pentation, results in a spectacular explosion of growth. The innocuous-looking $3\uparrow\uparrow\uparrow3 = 3\uparrow\uparrow3\uparrow\uparrow3 = 3\uparrow\uparrow7{,}625{,}597{,}484{,}987 = 3\uparrow3\uparrow3\uparrow3 \ldots \uparrow3$, which is a power tower with 7,625,597,484,987 threes. If a power tower of height 4 is enough to surpass the googolplex, think of what this can do. It's an unimaginably huge number, impossible to write out in a human lifetime *even in the form of a power tower.* If printed as a power tower, it would stretch all the way to the sun. Known as tritri, it is far larger than any number that we have mentioned up to this point and can barely even be comprehended by us mere mortals. Yet we have just begun the process. As large as tritri is, it's an insignificant dust mote in comparison to the great summit of Graham's number. Adding another up arrow brings us to $3\uparrow\uparrow\uparrow\uparrow3 = 3\uparrow\uparrow\uparrow3\uparrow\uparrow\uparrow3$ = $3\uparrow\uparrow\uparrow$tritri. Let's just go over what that means. Ascending up the power tower of threes, the first is 3, the second is $3\uparrow3\uparrow3$ = 7,625,597,484,987, and the third is $3\uparrow3\uparrow3\uparrow3 \ldots \uparrow3$ with 7,625,597,484,987 threes—in other words, tritri. The fourth power tower is $3\uparrow3\uparrow3\uparrow3 \ldots \uparrow3$ with tritri 3s, and so on. Then $3\uparrow\uparrow\uparrow\uparrow3$ is the tritrith power tower. This is a mind-bogglingly enormous step up from three up arrows. Yet it brings us only to G_1, the first of a series of G-numbers needed to reach the summit of Graham's number itself. Having arrived at base camp G_1, our next goal is G_2. Remember that every time we add a *single* extra up arrow, this produces a monstrous increase in the number being acted upon. Bearing this in mind, the definition of G_2 is $3\uparrow\uparrow\uparrow\uparrow \ldots \uparrow3$ *with G_1 up arrows.* Even grasping dimly what this means is enough to bring on a feeling of vertigo, a dizzying glimpse into how large numbers can be. Just one extra up arrow brings an awesome increase in size by everyday standards, yet G_2 has G_1 up arrows. And as you might guess, G_3 has G_2 up arrows, G_4 has G_3 up arrows, and so on. Graham's number, as it turns out, is G_{64}. The 1980 edition of the *Guinness Book of World Records* recognized it as the largest number ever used in a mathematical proof.

The problem that spawned Graham's number is fantastically hard to solve but quite easy to state. Graham was thinking about multidimensional cubes—hypercubes in *n* dimensions. Suppose any two corners, or vertices, are joined by a line that may be colored either red or blue. Graham asked, what's the smallest value of *n* so that for any such coloring, four vertices all lie in the same plane and all lines between any two of the vertices are the same color? Graham managed to prove that the lower limit for *n* was 6 and the upper limit was G_{64}. This colossal range reflects the difficulty of the problem. Graham could prove that a value for *n* satisfying the conditions existed but, as an upper bound, had to define a ridiculously huge number to prove anything. Since then mathematicians have managed to whittle down (relatively speaking) the range to values of *n* between 13 and $9\uparrow\uparrow\uparrow4$.

Graham's number, like the googol and googolplex, is a much-quoted example of a really large number. But it's also much misunderstood. First, it's no longer anywhere near the largest number ever defined. Second, in the search for ways of representing and defining new world record numbers, there's little point in starting from Graham's number and making elementary extensions of it.

In recent years, a branch of recreational math known as googology has sprung up, the sole aim of which is to push back the frontiers of truly big numbers by defining and naming ever-larger integers. Of course, anyone can think of a number larger than any that is stated. If I say "Graham's number," you might say "Graham's number plus 1," or "Graham's number to the power of Graham's number," or even "$G_{64}\uparrow\uparrow\uparrow\uparrow \ldots \uparrow G_{64}$ with G_{64} up arrows" (which is roughly G_{65}). However, all such extensions, involving repeated use of the same kind of operators, fail to bring about a radical change: the outcome is still Graham's numberish. In other words, it will be a number produced in roughly the same way as Graham's number itself, using a similar combination of tricks. Among serious googologists, an inelegant mishmash of existing numbers and functions that does little to enlarge the original large number is referred to as a "salad number" and is heavily frowned upon. Graham's number takes up-arrow notation and extends it to its limits. By contrast, salad numbers simply tack one insignificant operation onto Graham's number. What googologists want is not a naive and modest increase of it but an entirely new system that can be extended to the point where Graham's number appears utterly negligible. There is one such system that can be extended indefinitely. It's known as the fast-growing hierarchy, because of the prodigious growth rates it allows. What's more, it's a technique well tried and tested by mainstream mathematicians and so is often used nowadays as the benchmark for new ways of generating fantastically large numbers.

Two things are important to know from the start about the fast-growing hierarchy. The first is that it's a series of functions. A function in math is just a relationship, or a rule, for turning inputs into outputs. Think of a function as being like a little machine that transforms one value into another by always going through the same process. The process might be, for instance, "add 3 to the input." If we call the input x and the function $f(x)$, pronounced "f of x," then $f(x) = x + 3$.

The second key thing about the fast-growing hierarchy is that it uses ordinals to index the functions, which means how many times a process has to be carried out. We came across ordinals in the last chapter, on the subject of infinity. Ordinals, or ordinal numbers, tell us the position of something in a list or the order of something in a series. They can be finite or infinite. Everyone's familiar with finite ordinals, such as 5th, 8th, 123rd, and so on. But no one hears about the infinite ones unless they take a deeper plunge into math. It turns out that in trying to reach and define superlarge (but finite) numbers, both finite and infinite ordinals are extremely useful. Indexing functions with finite ordinals allows us to make our way to some reasonably large numbers. But the fast-growing hierarchy really comes into its own when it taps the power of infinite ordinals to control how many times a function is performed.

The starting point of the hierarchy is as straightforward as you could imagine. It's just a function that adds one to a number. Let's call this starting function f_0. So, if the number we want to put through the function mill is n, then $f_0(n) = n + 1$. This isn't going to get us to big numbers anytime soon—it's just counting up in steps of 1—so we'll move on to $f_1(n)$. This new function feeds the previous one into itself n times—in other words, $f_1(n) = f_0(f_0(\ldots f_0(n))) = n + 1 + 1 + 1 \ldots + 1$, with n ones, giving a total of $2n$. Again, this isn't too impressive in terms of how quickly it can get us into the land of giant numbers. But it reveals the process that ultimately gives the fast-growing hierarchy its immense power: recursion.

In art, music, language, computing, and math, recursion pops up in all kinds of different guises, but always it refers to something that feeds back into itself. In some cases, this just leads to an endless, repetitive loop. For example, there's the joke glossary entry "Recursion. See *Recursion*." On a more elaborate scale, a recursive loop appears in M. C. Escher's *Print Gallery* (1956), which shows a gallery in a city in which there's a picture of a gallery in a city in which . . . In engineering a classic example of recursion is feedback, where the output from a system gets routed back as input. It's a familiar problem for performers, such as rock musicians onstage, and often happens if a microphone is located in front of a loudspeaker to which it's connected. Sounds picked up by the mic come out of the speaker, having been amplified, then reenter the mic at a higher volume to be amplified again, and so it goes on until, very quickly, the familiar, ear-piercing squeal of feedback emerges. Recursion in math works along similar lines. A function takes the place of an electronic system, such as a microphone-amplifier-loudspeaker combination, and calls on itself, so that it feeds its own output back as input.

We had reached $f_1(n)$ on the ladder of the fast-growing hierarchy. The next rung, represented by $f_2(n)$, simply feeds $f_1(n)$ into itself n times. We can write it as $f_2(n) = f_1(f_1(\ldots f_1(n))) = n \times 2 \times 2 \times 2 \ldots \times 2$ with n 2s. This is the same as 2×2^n, which is an exponential function. Plugging in a value of, say, 100 for n, we get $f_2(100) = 100 \times 2^{100} = 126{,}765{,}060{,}022{,}822{,}940{,}149{,}670{,}320{,}537{,}600$, or about 127 billion trillion trillion. If this were a bank balance, it would be wealth far beyond the dreams of even Bill Gates, but it's a lot smaller than some of the numbers, such as a googol, that we've already run into. It's also smaller than the largest-ever lawsuit, for $2 undecillion ($2 trillion trillion trillion), filed on April 11, 2014, by Manhattan resident Anton Purisima after claims that he was bitten by a "rabies-infected" dog on a New York City bus. In a rambling twenty-two-page handwritten complaint, accompanied by a photo of an unreasonably outsize bandage around his middle finger, Purisima sued NYC Transit, LaGuardia Airport, Au Bon Pain (where he insisted he was routinely overcharged for coffee), Hoboken University Medical Center, and hundreds of others for more money than there is on the planet. His case was dismissed in May 2017 on the grounds that it "lacks an arguable basis either in law or in fact." Hopefully, Purisima's mathematical knowledge doesn't extend to the fast-growing hierarchy; otherwise, even larger suits (he has previously sued several major banks, Lang Lang International Music Foundation, and the People's Republic of China) may follow.

The function $f_3(n)$ involves n repetitions of $f_2(n)$ and leads to a number slightly greater than 2 to the power n to the n to the n . . . where the power tower is n high. This brings us to the stage of two up arrows, or tetration—the operation we came across earlier when making our assault on Graham's number. Continuing in the same vein, $f_4(n)$ involves three up arrows, $f_5(n)$ four up arrows, and so on, every increase of the ordinal by 1 having the effect of adding one more up arrow and boosting the number of up arrows to $n - 1$. This takes us into big-number territory by everyday—and even the litigious Anton Purisima's —standards. But just adding one up arrow at a time would never get us to Graham's number, let alone anything vastly bigger, in a sensible amount of time. For that we need to do something a little unexpected. To reach truly colossal finite numbers, we have to make use of numbers that are actually infinite.

The smallest kind of infinity, as we found in the last chapter, is aleph-null, the infinity of the natural numbers. Although aleph-null can't change in size—in other words, how much it contains—it can vary in length, depending on how its contents are organized. The shortest length of aleph-null is the infinite ordinal known as omega (ω). The next shortest is $\omega + 1$, then $\omega + 2$, $\omega + 3$, and so on, without end. These infinite ordinals, said to be countable because they can be put in a definite order, serve as the springboard for reaching some of the largest *finite* numbers ever conceived. To start off, we need a definition of what's meant by $f_\omega(n)$, the function indexed by the smallest infinite ordinal. We can't simply subtract 1 and apply the recursion process we talked about earlier because there's no such thing as $\omega-1$. Instead, we *define* $f_\omega(n)$ to be $f_n(n)$. Now, to be clear, we aren't saying that $\omega = n$. What we're doing is expressing $f_\omega(n)$ in terms of (finite) ordinals smaller than ω, so that we can reduce the function to a form that's useful for doing calculations. You might say we may as well write $f_n(n)$ instead of $f_\omega(n)$ and get the same result, but that would prevent the next crucial step—the step at which the immense power of the fast-growing hierarchy becomes apparent. As soon as we go from $f_\omega(n)$ to $f_{\omega+1}(n)$, something dramatic happens. Remember, when the ordinal that indexes the function goes up by 1, this feeds the previous function back into itself n times. If using a finite ordinal results in a fixed number of up arrows and using ω produces $n-1$ up arrows, then using $\omega + 1$ allows us to feed back into the number of up arrows n times, which amounts to a fantastic jump in the strength of the recursion.

To understand this, think about the function $f_{\omega+1}(2)$, which equals $f_\omega(f_\omega(2))$ using our recursion rule. Because we defined $f_\omega(2)$ to be the same as $f_n(2)$, we can rewrite $f_{\omega+1}(2)$ as $f_\omega(f_\omega(2))$, just replacing the innermost ω with 2. (We can't work out the value of the outer f_ω until we know what value the inner one will take.) As it turns out, $f_2(2) = 8$, so now we're left with $f_{\omega+1}(2)$, which equals $f_\omega(8)$. Finally, we can simplify the outermost ω and get $f_8(8)$, which is a number with seven up arrows. While this shows how $f_{\omega+1}$ can be used to feed back into the number of up arrows, it doesn't give a clear impression of the function's awesome capability. This becomes apparent only as n gets larger and the corresponding number of feedback loops grows. Putting $n = 64$ gives $f_{\omega+1}(64)$, which is approximately Graham's number. The next step up the fast-growing hierarchy, $f_{\omega+2}(n)$, breaks into new territory because it plugs all of the mathematical machinery used to reach the level of Graham's number back into itself. The result is a number we can write roughly as $G_{G \ldots 64}$ (with sixty-four levels of the G subscript), although there's no hope of trying to grasp, even vaguely, what this means.

The countably infinite ordinals stretch away into the far distance, each successive one providing the basis for a recursive function that utterly dwarfs the power of the one before it. The omegas alone form a sequence so long that it ends only when we reach omega raised to a power tower of omega omegas. This mighty ordinal, known as epsilon-zero, is so vast that it can't be described within our conventional system of arithmetic, known as Peano arithmetic. With each step along the eternal road of omegas, the finite number generated by recursion increases by an amount that defies comprehension. But beyond the loftiest omegan power tower lie tier upon higher tier of yet greater infinite ordinals—first the epsilons, then the zetas, and on and on, without end—as we saw earlier in our exploration of infinity. These increasingly vast ordinals serve to inform more and more powerful degrees of feedback. Finally, we reach a tremendously large ordinal known as Gamma-zero (Γ_0) or, more magnificently, the Feferman-Schütte ordinal, after American philosopher and logician Solomon Feferman and German mathematician Karl Schütte, who first defined it. While Gamma-zero is still countable, and there are countable ordinals beyond it, to actually define it requires the use of uncountable ordinals (ones that can't be made from rearranging elements of aleph-null but instead require aleph-one or more elements). This process is reminiscent of how the fast-growing hierarchy itself evolves. Just as we have to resort to using infinite ordinals in the fast-growing hierarchy to describe huge finite numbers, so we need to turn to uncountable ordinals to describe truly tremendous countably infinite ordinals. There aren't any adjectives left to describe adequately the size of the finite numbers to which the Feferman-Schütte ordinal, and others beyond it, gives rise by recursion. Nor does any mathematician have a brain big or clever enough to grasp the immensity of the numbers their recursive techniques can spawn. But that doesn't stop them from coming up with even mightier methods for big-number generation. Notable among these is the TREE function.

As the name suggests, a tree in mathematics can have the appearance of a tree that grows in the ground or a family tree, with branches spreading out from a common trunk. Mathematical trees are a special variety of what in math are known as graphs. Usually we think of graphs as being charts drawn on graph paper, in which one value is plotted against another. But the kinds of graphs we're talking about in connection with trees are different: they're ways of representing data in which dots, called nodes, are connected by line segments, called edges. If it's possible to start at a node, move to other nodes along edges, and then return to the starting node without repeating any edges or nodes, then the route taken is known as a cycle and the graph is said to be cyclic. If it's possible to start at any node and travel to any other node, without repeating an edge or node along the way, then the route taken is called a path and the graph is said to be connected. A tree is defined as a graph that is connected but has no cycles. Both family trees and biological trees also have this kind of structure. If a unique number or color is assigned to each node, then the tree is said to be labeled. Furthermore, if we assign one node to be the root, then we have a rooted tree. One useful property of rooted trees is that for any node, we can always trace back a path to the root.

Some mathematical trees that have the same kind of branching structure as a real tree can be fitted inside others of their kind. They're said to be homeomorphically embeddable, which is a fancy way of saying they're similar in form or appearance and one of them is like a smaller version of the other. Of course, mathematicians are a bit more precise about the definition. They start with a larger tree and see how much of it can be pruned using a couple of different methods. First, if there's a node (except for the root node) that has just two edges leading into, or from, it, that node can be removed and the two edges joined into one. Second, if two nodes are joined by a single edge, the edge can be contracted and the two nodes compressed into one. The color of this new node is the color of whichever node was originally closer to the root. If a smaller tree can be made by applying these two steps in any order to the larger tree, the smaller one is said to be homeomorphically embeddable in the larger. American mathematician and statistician Joseph Kruskal proved an important theorem to do with this kind of tree. Suppose there's a sequence of them so that the first tree can have only one node, the second up to two nodes, the third up to three, and so on, and that no tree is homeomorphically embeddable in any subsequent tree. What Kruskal found is that such a sequence always has to end at some point. The question is how long can the sequence be?

In response, American mathematician and logician Harvey Friedman, listed in the *Guinness Book of Records* in 1967 as the world's youngest professor (an assistant professor at Stanford, aged just eighteen), defined the tree function, TREE(n), as the maximal length of such a sequence. Friedman then investigated the output of the function for different values of n. The first tree consists of a single node of a certain color, which can't be used again. If $n = 1$, this is the only color and the sequence immediately stops, so that TREE(1) = 1. If $n = 2$, there's one more color. The second tree can contain up to two nodes, so it contains two nodes, both of this color. The third tree also must contain only this color, but can only have one node, as otherwise the second tree would be homeomorphically embeddable in the third. Beyond that, no more trees are possible, so TREE(2) = 3. The big shock, as Friedman found, comes when we reach TREE(3). In a sudden explosion of complexity and proliferation, the number of nodes far surpasses Graham's number and reaches the small Veblen ordinal, that extraordinarily unsmall number we mentioned in our travels among the various infinities in the fast-growing hierarchy.

So popular has googology—the quest to define ever-larger numbers—become that it's given rise to several contests. One of the first was the Bignum Bakeoff, organized by American math whiz kid David Moews in 2001. Competitors in the Bakeoff were challenged to produce the biggest number they possibly could from a computer program that was no more than 512 characters long (ignoring spaces) in the programming language C. No present-day computer could actually complete any of the programs submitted within the lifetime of the universe, so the entries were analyzed by hand and ordered based on their position in the fast-growing hierarchy. The winning entry was a program called loader.c, after its creator, New Zealander Ralph Loader. A computer with an unfeasibly large memory and an outlandish length of time available would be needed to generate the final output. But if it could be done, the result would be Loader's number—an integer known to be bigger than TREE(3) and some other heroic inhabitants of the googologist's cosmos, such as SCG(13), the thirteenth member of a sequence known as subcubic graph numbers (similar to the TREE sequence, but consisting of graphs in which each vertex has at most three edges).

In 2007 a large-number contest called the Big Number Duel pitted philosophers and old graduate school friends Agustin Rayo (a.k.a. the Mexican Multiplier) of MIT against Adam Elga (a.k.a. Dr. Evil) of Princeton in a back-and-forth tournament to see who could define the most colossal integer. The numerical slugfest, which blended comedy, the melodrama of a world-title boxing match, and convoluted mathematical, logical, and philosophical maneuverings, took place in a packed room in MIT's Stata Center. Elga opened optimistically with the number 1, perhaps hoping that Rayo would have an off day. But Rayo swiftly countered by filling the entire blackboard

The poster for MIT's Big Number Duel. (ADAM ELGA)

with 1s. Elga promptly rubbed out a line near the bottom of all but two 1s, turning them into factorial signs. Then the duel progressed, eventually transcending the bounds of familiar mathematics, until each competitor was inventing their own notations for ever-larger numbers. It's been reported that, at one point, a spectator asked Elga, "Is this number even computable?" to which Elga, after a brief pause, replied, "No." Finally, Rayo delivered the knock-out blow with a number that he described as "the smallest positive integer bigger than any finite positive integer named by an expression in the language of first-order set theory [FOST] with a googol symbols or less." Just how large Rayo's number is, we don't know and probably never will. No computer could ever calculate it, even given access to a universe that could hold a googol symbols or more. The issue isn't having enough time and space; Rayo's number is uncomputable in the same way that the halting problem is uncomputable.

For now, when it comes to the largest integers that we can sensibly talk about, Rayo's number more or less marks the boundary with the unknown. Some bigger numbers have been named, notably BIG FOOT, which was announced in 2014. But gaining even a hazy understanding of BIG FOOT would mean entering a strange domain known as the oodleverse and learning the language of first-order oodle theory—a venture best tackled with a higher-math degree and a wry sense of humor. In any case, the largest named numbers to date all build on the same kind of concepts used to reach Rayo's number.

To penetrate further into endless number space, googologists must build on old methods or develop new ones, just as sending spacecraft deeper into physical space depends on breakthroughs, great and small, in propulsion technology. For the present, big-number seekers will probably need to rely on the same tricks used by Rayo but apply them to beefed-up versions of first-order set theory. They might, for instance, add axioms that would give FOST access to even more staggeringly huge infinities, which could then be used to generate new record-breaking finite numbers.

To be frank, most professional mathematicians aren't too fussed with the subject of gigantic numbers just for the sake of defining them, any more than they are with extending the known digits of pi. Googology is a sideshow, a bit of intellectual machoism, NASCAR for the number theorist. At the same time, it isn't without its merits. It does expose the limits of our current mathematical universe, just as peering into space with the world's largest telescopes pushes back the frontiers of the physical cosmos.

It's tempting to think that such huge numbers as Rayo's bring us closer to the infinite. But in fact, that isn't the case. Infinite numbers may be used to generate finite ones, but, however high we go, there's never a point at which the finite merges with the infinite. The truth is that seeking out ever-larger finite numbers gets us no nearer to infinity than the "1, 2, 3" of our early childhood.

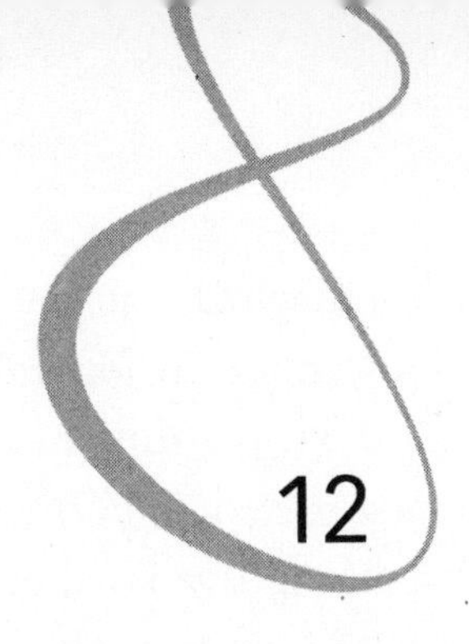

12

BEND IT, STRETCH IT, ANY WAY YOU WANT TO

A child['s] . . . first geometrical discoveries are topological. . . . If you ask him to copy a square or a triangle, he draws a closed circle.

—JEAN PIAGET

Topology is precisely the mathematical discipline that allows the passage from local to global.

—RENÉ THOM

There's an old joke that asks, what is a topologist? The answer: someone who can't tell the difference between a doughnut and a coffee cup—or, more to the point, who doesn't care about the difference. In topology a doughnut and a coffee-cup shape are equivalent because (assuming they were made of something like clay) the one could be gradually deformed into the other: the handle deformed into the hole in the doughnut and the rest of the coffee cup gradually morphed into the ring around it. "Hole" here has a specific meaning. A hole in topology must have two ends and pass all the way through, as it does in the case of a doughnut shape or, to give it its formal name, a torus. What sometimes in everyday language is called a hole, such as one dug in the ground, isn't a hole to a topologist because it doesn't have two openings and can be gradually deformed until it's completely filled in. In a nutshell, then, topology is the study of properties that stay the same if something changes in shape without having a hole put in it or being cut. It's a modern extension of geometry that gives rise to a lot of weird results and pops up in all kinds of unexpected places.

The 2016 Nobel Prize in Physics was awarded to a trio of British scientists, Duncan Haldane, Michael Kosterlitz, and David Thouless, for their work on so-called exotic states of matter. Under certain conditions, such as very low temperatures, a material can undergo a sudden, unexpected switch in behavior. One morning in February 1980, a German physicist, Klaus von Klitzing, was experimenting with a supercooled, ultrathin sliver of silicon in a powerful magnetic field when he noticed something bizarre. The silicon had begun conducting electricity only in packets of certain sizes: a smallest-size packet, a packet exactly twice as big, three times as big, and so on, or nothing at all—there were no in-between amounts as you get with an ordinary electrical current. This phenomenon is known as the quantum Hall effect, and von Klitzing won the 1985 Nobel Prize in Physics for shedding new light on it. Evidently, the silicon had jumped into some new physical state, in which, as happens whenever there's a change of state, there must have been a rearrangement of atoms. Theorists, however, struggled to explain how such a rearrangement could take place in a layer of silicon so thin that the atoms within in it had no room to move up or down. Then Kosterlitz and Thouless came up with a novel idea. As the silicon cooled, they suggested, swirling pairs of silicon atoms formed and then spontaneously separated into two miniature vortices at the critical temperature of the transition. Thouless set to work on the math behind these spinning transitions and found that it could best be formulated in terms of topology. The electrons in the material undergoing the change were forming what's known as a topological quantum fluid, a state in which they flowed collectively only in whole numbers of steps. Working independently, Haldane found that these fluids can spontaneously appear in ultrathin layers of semiconductors even in the absence of strong magnetic fields.

After the announcement of the 2016 prize in Stockholm, a member of the Nobel committee stood up and drew from a paper bag a cinnamon bun, a bagel, and a (Swedish) pretzel. These were different, he pointed out, in a number of ways—in flavor, for instance, sweet or salty, and general appearance. But to a topologist, only one of their differences mattered: the number of holes—zero in the bun, one in the bagel, two in the pretzel. The winners of the prize, he explained, had found a way to link the sudden appearance of exotic physical states to changes in topology—effectively, the "hole-iness" of the underlying abstract structures. In doing so, they had found a new and tremendously important application for a subject that has spawned some of the most surprising results in mathematics.

Take two prints of the same picture. Put one of them down flat on a table, and then crumple up the other, any way you like, providing you don't actually tear it, and place it somewhere on top of the uncrumpled print. It's an inescapable fact that at least one point on the crumpled copy will lie directly over the corresponding point on the flat picture. (Strictly speaking, the math we're talking about here deals with continuous quantities, whereas in the real world matter is grainy because it's made of atoms and so forth, but the result is still valid to a very good approximation.) The same thing is true in three dimensions, so if you stir a glass of water, for however long you like, at least one water molecule will be in the same position after stirring as before. The first mathematician to publish a proof of this was Dutchman Luitzen Brouwer, in the early years of the twentieth century, and so it became known as Brouwer's fixed-point theorem.

Another curious result first proved by Brouwer, in 1912, though it had been proposed earlier by prolific French mathematician Henri Poincaré, is the hairy-ball theorem. This states that no matter how much you comb a ball entirely covered with hair, it's impossible to make the hair lie flat at every point: somewhere it must be sticking straight up. Brouwer (and Poincaré) didn't actually talk about hairy balls, but did discuss less evocative stuff about continuous vector fields tangent to a sphere that must have at least one point where the vector is zero (at right angles to the sphere). But it amounts to the same thing. In more practical terms, because the velocity of wind along earth's surface is a vector field, the theorem guarantees that there has to be somewhere on the planet where the wind isn't blowing. Another truism, closely related to the fixed-point theorem and known as the Borsuk-Ulam theorem, has something else to say about meteorological conditions: at any given moment, there exist two points on opposite sides of the planet with exactly the same temperature and barometric pressure. You might say that this is likely to happen quite a lot just by chance, but the Borsuk-Ulam theorem guarantees mathematically that it's bound to be the case.

Yet another strange-but-true fact follows from the Borsuk-Ulam theorem—the so-called ham-sandwich theorem. Make a sandwich from ham and cheese, and this result says that it's always possible to slice the sandwich in such a way that the two parts have an equal amount of bread, cheese, and ham. In fact, the three ingredients don't even have to be touching: the bread could be in the bread box, the cheese in the fridge, and the ham on the kitchen counter. Or, for that matter, they could be in different parts of the galaxy. There's always going to be one flat slice (in other words, a plane) that bisects each one of them.

All these strange theorems—fixed point, hairy ball, Borsuk-Ulam, and ham sandwich—have sprung from the fertile ground of topology, *tópos* being the Greek for "place" or "locality." It's a subject that, in daily life, we don't generally hear much about. Everyone's familiar with geometry—the subject, with ancient origins, that deals with the shape, size, and relative position of figures such as triangles, ellipses, pyramids, and spheres. Topology is related to geometry and also to set theory, and, as we've mentioned, it has to do with properties that stay the same even when a figure is bent or stretched out of shape—properties known as topological invariants. Examples of such invariants include the number of dimensions involved and the connectedness or how many separate pieces of which something is composed.

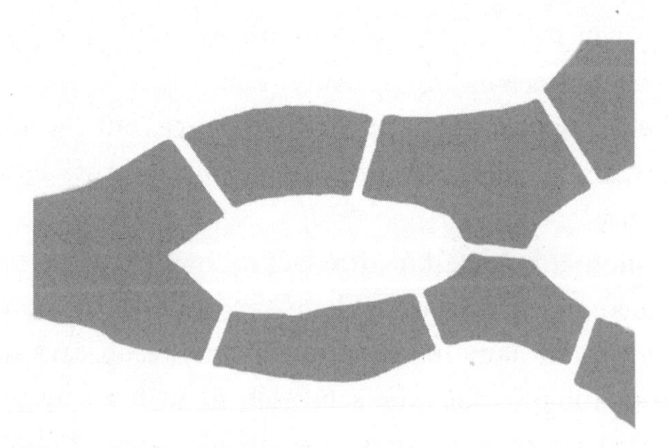

The seven bridges of Königsberg across the River Pregel. (DAVID DARLING)

The origins of topology can be traced back to the seventeenth century when German polymath Gottfried Leibniz raised the possibility of dividing geometry into two parts: *geometria situs*, the geometry of place, and *analysis situs*, the analysis or taking apart of place. The former, which pretty much covers the geometry we learn about in school, deals with familiar concepts such as angles, lengths, and shapes, while the latter is concerned with abstract structures that are independent of those concepts. Swiss mathematician Leonhard Euler subsequently published one of the first papers on topology, in which he showed that it was impossible to find a way around the old seaport city of Königsberg in Prussia (now Kaliningrad in Russia) that would cross each of its seven bridges exactly once. The result didn't depend on measurements, such as the lengths of the bridges or how far they were apart, but only on how they were connected with the land, either with islands in the river or with the riverbanks. He found a general rule for solving such types of problems and in doing so gave birth to a new field of study, within topology, called graph theory.

Euler also discovered a now-famous formula for polyhedra (3-D solids with flat polygonal faces): $v - e + f = 2$, where v is the number of vertices (corners), e the number of edges, and f the number of faces. Again, the result is topological because it involves properties of geometric forms that don't depend on measurements.

Another pioneer in the field was August Möbius, with his explorations of a band with a half twist that now bears his name, though his compatriot Johann Listing published his own findings on the band a few years before Möbius, in 1861. If a strip of paper is twisted through 180 degrees and then has its

A Möbius band, a shape with only one "side" when embedded in 3-D. (DAVID BENBENNICK)

ends glued together, the result is a shape with only one surface—a fact easily demonstrated by drawing a pencil line all the way around the middle of the band until it joins up with the starting point. The act of joining the ends of the band following a half twist makes a Möbius band a different beast in the eyes of a topologist from an ordinary band or open-ended cylinder. Whenever a shape is torn, or the ends of it joined together, it becomes topologically something new. This fact leads to another feature of topology: it is well suited to describing sudden jumps in the state of a system, as the winners of the 2016 Nobel Physics Prize discovered.

In ordinary geometry, all figures are treated as being rigid and noninterchangeable. A square is always a square, a triangle always a triangle, and the one can never mutate into the other. Straight lines must stay perfectly straight, and curves remain curves. However, in topology, shapes are allowed to lose their structure and become flexible while remaining essentially the same, providing they aren't cut at any points or have separate parts joined together. For example, a square can be stretched and deformed until it becomes a triangle yet in topological terms be unchanged: they're said to be homeomorphisms. Likewise, both are identical to discs (circles with a solid interior). In three dimensions, a cube is homeomorphic to a ball (a sphere with a solid interior). In other words, the surface of a cube is topologically identical to that of a sphere. However, a torus or doughnut shape is fundamentally different from a sphere, and no amount of stretching will make them the same.

The number of holes in a shape is known as its genus. So a sphere and a cube have genus 0, an ordinary doughnut shape genus 1, a two-holed torus genus 2, and so on. Three-dimensional topology can also take into account more complex factors, such as the structure of the surrounding space, which is what allows knots to form. Confusingly, in knot theory, most knots that we learn to tie are not knots at all. A mathematical knot differs from a knot in, say, a shoelace or a rope because its ends are joined together, so that it can't be undone.

One way to think of a true knot is as a circle or any other closed loop inhabiting 3-D Euclidean space. No amount of stretching or twisting will serve to untie it. The only way to make a true (mathematical) knot from a piece of string is to join the ends by, for example, taping them together. Using this method, the simplest knot is the unknot, which is just a plain loop. After this, things get more complicated.

The simplest nontrivial knot is the trefoil knot, which is the kind that people often make if you ask them to tie a knot in a piece of string and if the ends are then joined together. More complex is the figure-eight knot or those formed from a combination of several basic knots. Two common examples are the square knot, also known as the reef knot, and the granny knot, both formed from two trefoil knots.

The first person to take an interest in knots from a mathematical point of view was German mathematician Carl Gauss in the 1830s. He came up with a way to calculate the linking number—a number that, in the case of two closed curves in 3-D, tells how many times the curves wind around each other. Links, like knots, have a central place in topology. Mathematical knots and links crop up in nature too, in electromagnetism and quantum mechanics, for instance, and in biochemistry.

Just as there's an unknot, there's an unlink, which is just two separate circles that aren't joined in any way. Knots are simple links consisting of a single circle, but more complex links are possible by adding more circles. The Hopf link, consisting of two circles linked together once, is named after German topologist Heinz Hopf, although Gauss had studied it a century earlier and it has long figured in artwork and symbolism. The Buzan-ha, a Japanese Buddhist sect founded in the sixteenth century, used it in its crest. More interesting are the Borromean rings, which employ three circles. What makes these unusual, and at first sight seemingly impossible, is that although no pair of circles is linked in any way, all three are. This means that if any of the three rings are removed, no matter which one is chosen, the other two can then be easily separated. The name comes from the Italian noble family of Borromeo, who used this link as part of their coat of arms, but the symbol dates back to antiquity. It crops up in Viking artifacts in the form of three interlocking triangles known as the Walknot (meaning knot of the slain) or Odin's triangle. The motif also appears in various religious contexts, including old Christian church decorations, where it symbolizes the holy Trinity.

Knots and links have been found in the chemistry of life itself. Proteins are well known for their ability to fold into certain specific shapes, which are crucial to the way they function in biological systems. What came as a surprise to biologists was the discovery, starting in the mid-1990s, that they can fold to become knotted and may even form linked rings. Knots on an everyday scale call for some kind of threading to be done intentionally. It was hard to see how a protein could spontaneously self-assemble and, at the same time, manage to put a knot in itself. In fact, most mathematical models used to predict the outcome of protein folding, based on energy considerations, explicitly threw out any structures that were knotted because they were thought to be impossible. It's an ongoing issue for researchers to understand how knotted proteins fold—and why.

At the start of 2017, a team of chemists from the University of Manchester announced that it had created the tightest knot ever seen. Made from 192 atoms linked in a chain, the knot was a mere 2 millionths of a millimeter wide—about 200,000 times thinner than a human hair. The atoms, of carbon, nitrogen, and oxygen, formed a strand that crossed itself eight times and curled around into a circular triple helix. Between each crossing point, the distance that defines the tightness of the knot, were just 24 atoms.

Other unexpected topologies have been found in the scientific world, one of the most surprising of which is the Möbius band, mentioned earlier. In 2012 chemists at the University of Glasgow reported that they had turned a symmetric, ring-shaped molecule into an asymmetric one by adding a molybdenum oxygen unit, with the formula Mo_4O_8, into the ring. The effect of introducing the new unit was to give the ring a half twist, thereby producing a Möbius topology.

Making a Möbius band can literally be child's play. But this isn't the case with another one-sided surface known as the Klein bottle, after the German mathematician who first described it, Felix Klein. It may originally have been called *Kleinsche Fläche*, meaning the Klein surface, which then got passed along erroneously as *Kleinsche Flasche*, the Klein bottle. In any event, the latter name has stuck and may have

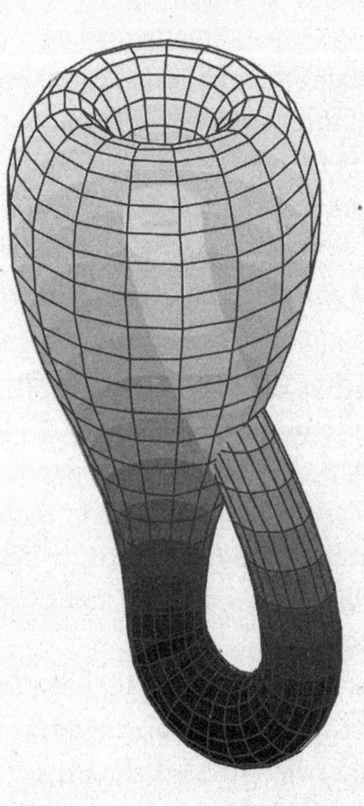

A Klein bottle immersed in three dimensions. The "inside" and "outside" are, in fact, the same side. This cannot normally be achieved (there is no embedding into 3-D), so the Klein bottle must intersect itself.

helped give the object wider recognition, despite "surface" being a better description.

Unlike the Möbius band, the Klein bottle has no edges or boundaries, a property it shares in common with the sphere. But unlike a sphere, the Klein bottle doesn't have an inside and an outside—the two are identical—because there's just a single surface folded back on itself. We're not used to this kind of thing. In the real universe, we're accustomed to objects, such as bubbles, boxes, and bottles of Beaujolais, having a well-defined interior and exterior, so that they enclose a certain volume of space. But because the Klein bottle doesn't separate space into two different regions, it encloses nothing and therefore has zero volume.

Spheres, tori, and Möbius bands are all examples of two-dimensional surfaces that can be "embedded" in three-dimensional space. Embedding has a precise mathematical meaning, but in everyday terms it can be thought of as sticking one space inside another, different, space. It's important to keep in mind that spheres, Möbius bands, Klein bottles, and other geometrical objects are abstractions with properties that don't depend on the nature of the space they're put into—how many dimensions it has, whether it's flat or curved, and so on. But certain things about them do change from one embedding to another. For example, a torus can be embedded in 3-D, which is how we normally encounter it, and it then appears with a hole—a true mathematical hole—and an inside and an outside.

Some readers may be old enough to remember the classic arcade game Asteroids. In this, the player controls a spaceship and tries to knock out wayward asteroids and occasional flying saucers that pass by. At first glance, this seems to have nothing at all in common with the familiar doughnut-shaped ring torus. However, topologically, they are one and the same—both are toroidal. The hole in a doughnut is a feature caused by the embedding of a torus in 3-D and isn't an inherent property of all tori. In Asteroids space, the underlying toroidal topology manifests itself not as a hole but as the ability of things that disappear off one side of the screen to immediately reappear on the opposite side. A torus can also be embedded in 4-D, and one of the possible outcomes of this is the Clifford torus, named after Victorian mathematician William Kingdon Clifford, who was also the first person to suggest that gravitation might be an effect of the geometry of the space in which we live. Unlike the ring torus that we know well, with its clearly defined interior and exterior, the Clifford torus doesn't divide space and so can't be said to have an inside and an outside.

The same is true of the Klein bottle. Austrian Canadian mathematician Leo Moser described, in the form of a limerick, how the original idea for this shape came about:

A mathematician named Klein
Thought the Möbius band was divine.
Said he: "If you glue
The edges of two,
You'll get a weird bottle like mine."

This is why the Klein bottle has no edges—when the edges of two Möbius bands (a left and a right) are brought together, they form one continuous surface that is smoothly connected at all points. Another way to make a Klein bottle is to start with a rectangle, join one pair of opposite sides to make a cylinder, and then join the other pair of sides after making a half twist. This second step, simple as it sounds, is actually impossible in three dimensions. Access to a fourth is needed so that the surface can be made to pass through itself without a hole. That little difficulty doesn't stop people from making 3-D models of

Klein bottles that are nearly but not quite accurate representations. Notable experts of the art are Clifford Stoll, of Oakland, California, who runs the Acme Klein Bottle company, and Alan Bennett, of Bedford, England, who fashioned a series of Klein bottles analogous to Möbius bands, with odd numbers of twists greater than one, for the Science Museum in London. What these craftsmen have created are known by mathematicians as 3-D "immersions" of Klein bottles. The distinction between an immersion and an embedding is a technical one, but what it boils down to is that a 3-D model (an immersion) of a Klein bottle will always have a point of self-intersection where the surface passes through itself. A true Klein bottle has no such self-intersection, and, indeed, none is present in a 4-D embedding of it.

Another important feature of the Klein bottle, and of any surface, is its orientability. Most surfaces we meet in the physical world are said to be orientable. What this means is that if you were to draw a small circular arrow on the surface, pointing either clockwise or counterclockwise, and then slide the arrow all the way around the surface and back again to where it started, it would still point in the same direction. This would happen on a sphere or torus, for example, so these are orientable surfaces. But try the same experiment with a Klein bottle or a Möbius band, and the arrow would have reversed its direction because these are nonorientable surfaces.

Topologists spend a lot of time flitting, in their mind's eye, between spaces of different dimensions. So they've come up with their own vocabulary to be able to generalize about things as they do this dimension hopping. "Embedding" and "immersion" are a couple of terms used in this respect; another is "manifold," which is a generalization of the term "surface" to other dimensions. By definition, surfaces are two-dimensional, so rather than say "2-D surface" (which is tautological), we should really say "2-D manifold." Spheres, tori, Möbius bands, and Klein bottles are all examples of 2-D manifolds. The first three of these can be embedded in 3-D but not the Klein bottle. Lines and circles are 1-D manifolds, and, although we can't visualize them properly, there are 3-D manifolds, 4-D manifolds, and so on. One of the simplest 3-D manifolds is the 3-sphere. Just as an ordinary sphere, or 2-sphere, is a surface that forms the boundary of a ball in three dimensions, a 3-sphere is an object with three dimensions that forms the boundary of a ball in four dimensions. We can't accurately imagine what the 3-D analogue of a surface might look like, let alone boundaries in even larger numbers of dimensions. But despite this handicap, mathematicians have all the tools they need to deal with them.

Working in higher dimensions springs some surprises. In 4-D, for instance, circles can't be linked and ordinary knots don't exist. The same is true in all higher dimensions. Something very bizarre happens in four-dimensional space: spheres themselves can become knotted. We can't visualize this, but then the idea of circles being knotted without self-intersecting would be impossible for two-dimensional beings to imagine.

Like all other areas of mathematics, topology is a dynamic subject in which new discoveries are being made every year, and problems, old and new, remain to be solved. One of the most important notions in topology, and in math as a whole, is known as Poincaré's conjecture. It isn't important because of any obvious practical applications. It won't, so far as we know, help us get to Mars faster or find a cure for aging. Its interest to mathematicians is purely theoretical, as part of the effort to classify higher-dimensional surfaces, or manifolds.

The conjecture was first put forward in 1900 by Henri Poincaré, one of the founders of topology as a precise discipline and regarded by some as the "last universalist," in that he was an expert in all the areas of math as they existed during his lifetime. Poincaré came up with a technique called homology, which, loosely speaking, is a way of defining and categorizing holes in manifolds. This isn't as straightforward as it may sound, because mathematical holes can be sneaky things that aren't as easy to spot and count as the ones in, say, a pretzel or an old sock. The two-dimensional space in Asteroids, for example, is topologically equivalent to a torus, although a torus appears to clearly have a hole, whereas the Asteroids space doesn't seem to have any. Keep in mind that mathematical holes are abstract things that can be harder to imagine than, say, a hole in a doughnut and also that they are surrounded by "loops," so that homology can also be defined as a way of analyzing the different types of loops in manifolds.

Poincaré's original conjecture was that homology was enough to tell whether any given three-dimensional manifold was topologically equivalent to a 3-sphere. However, within a few years, he himself disproved this by finding the Poincaré homology sphere, which isn't a true 3-sphere but has the same homology as it. After further research, he restated his conjecture in a new form. In plain language, it says that any finite three-dimensional space, providing it has no holes in it, can be continuously deformed into a 3-sphere. Despite much effort during the twentieth century, the conjecture went unproven. So important was it considered to be that, in 2000, the Clay Mathematics Institute listed it as one of seven major problems for which a solution would win a $1 million prize. Three years later, the conjecture was proved correct by Russian mathematician Grigori Perelman as a consequence of his proof of a closely related problem called Thurston's geometrization conjecture.

In 2005 Perelman was awarded the Fields Medal—arguably the most prestigious honor in mathematics and often considered to be equal in stature to a Nobel Prize. Then, in 2010, came the announcement that he had met the criteria for the $1 million Clay Institute Prize. However, he turned down both these awards, apparently on ethical grounds. First, in his view, they didn't acknowledge the important contributions of others, notably American mathematician Richard Hamilton, whose work Perelman had built upon. He was also unhappy about what he considered to be a lack of good conduct by some researchers, especially Chinese mathematicians Zhu Xiping and Huai-Dong Cao, who, in 2006, published a verification of the Hamilton-Perelman proof but seemed to imply that the proof was actually their own work. They later retracted their original paper, titled "A Complete Proof of the Poincaré and Geometrization Conjectures: Application of the Hamilton-Perelman Theory of the Ricci Flow," and issued another with more modest claims. But the damage had been done as far as Perelman was concerned: he was dismayed by their behavior and also by the lack of criticism of it by others in the field. In an interview with the *New Yorker* in 2012, he said, "As long as I was not conspicuous, I had a choice. Either to make some ugly thing [a fuss about the ethical breaches he perceived] or, if I didn't do this kind of thing, to be treated as a pet. Now, when I become a very conspicuous person, I cannot stay a pet and say nothing. That is why I had to quit." Whether Perelman has now permanently retired from mathematics or is quietly working on other problems isn't clear. Certainly, he is not one to enjoy the limelight. "I'm not interested in money or fame," he said after being awarded the Clay Institute Prize. "I don't want to be on display like an animal in a zoo." His place in history, however, is ensured, having finally settled one of the most important and difficult questions in topology.

Another well-known thorn in the side of topologists goes by the name of the triangulation conjecture, and it, too, has recently been resolved—but this time in the form of a disproof. In plain English, the issue is whether every geometric space can be divided into smaller pieces; the triangulation conjecture proposes that it can. In the case of a sphere, for instance, it's possible to completely tile its surface with triangles. A regular icosahedron—a polyhedron with twenty sides made of equilateral triangles—is a rough approximation of a sphere, but we can improve on this indefinitely by using as many triangles as we like and of any shape. A torus can be "triangulated" in the same way. A three-dimensional space can be sliced up into an arbitrary number of tetrahedrons. But is it possible to triangulate geometric objects in all higher dimensions with higher-dimensional equivalents of a triangle? In 2015 Ciprian Manolescu, a Romanian mathematics professor at the University of California at Los Angeles, managed to prove that it isn't. Manolescu, a child prodigy who is the only person ever to have racked up three consecutive perfect scores in the International Mathematics Olympiad, first encountered the triangulation problem as a graduate student at Harvard in the early 2000s. At the time, he dismissed it as "an unapproachable problem," but years later realized that a theory he had written about in his PhD dissertation, concerning something called Floer homology, was the very thing needed to resolve the issue. By bringing his earlier work to bear, he managed to show that there are some seven-dimensional manifolds that have no triangulation, thereby disproving the triangulation conjecture. It was a remarkable feat given that, using other methods, even spaces of four dimensions are still too complex to analyze with respect to their triangulation.

In the early 1980s, American geometer William Thurston, who died in 2012, envisioned a project that would identify every three-dimensional manifold. In two dimensions this had already been done. The 2-manifolds are the sphere, torus, two-holed torus, three-holed torus, and so on. To these we can add nonorientable surfaces, such as the Klein bottle and the projective plane (made by joining two Möbius bands of the same handedness along their edges). Thurston used a technique that allowed many of these 2-manifolds to be represented by polygons. For example, if you take a square and join opposite edges, the result is a torus. The two-holed torus is tougher to produce, but Thurston found a way. He represented a two-holed torus by joining certain pairs of edges of an octagon, which is embedded in the hyperbolic plane. This embedding avoids a difficulty that crops up if the octagon is Euclidean. In this case, the two-holed torus would have a single point common to all vertices of the octagon, which would have an angle sum of 1,080 degrees and not 360, as is required. In hyperbolic geometry—geometry on saddle-shaped surfaces, or, more precisely, ones that curve the opposite way to a sphere and at a constant rate—octagons of the correct size can have angles of 45 degrees, thereby fixing the problem.

Thurston tried to do a similar thing in three dimensions. In 2-D there are three types of uniform geometry: elliptic, Euclidean, and hyperbolic. The elliptic and Euclidean geometries can easily be embedded in space, but hyperbolic geometry can't, which is why it wasn't discovered until much later. In 3-D all three of these geometries have an analogue, but there are also some others, for a total of eight geometries. Of these, the hyperbolic one is the most complex and hardest to work with, just as it is in 2-D. In 2012 Ian Agol managed to enumerate all of the hyperbolic manifolds (then the only unsolved case). His methods involved techniques that at first sight appeared to bear no relation to the original problem, such as using complexes made of cubes of various dimensions and analyzing the hyperplanes that bisect these cubes. These manifolds have real applications. For instance, some cosmologists have suggested that the geometry of the universe as a whole is elliptic and is a finite manifold, having the structure of a dodecahedron with certain faces identified. This manifold can be classified by Agol's techniques.

Of course, there are still many unsolved problems in topology, and perhaps there always will be, given that as the boundaries of the known are pushed back, they reveal more of the extent of our ignorance. But topology is no longer the specialized and seemingly impractical subject it was a century or more ago. It has countless real-world applications, including robotics, condensed-matter physics, and quantum field theory, and its ideas can be found in almost all areas of mathematics today.

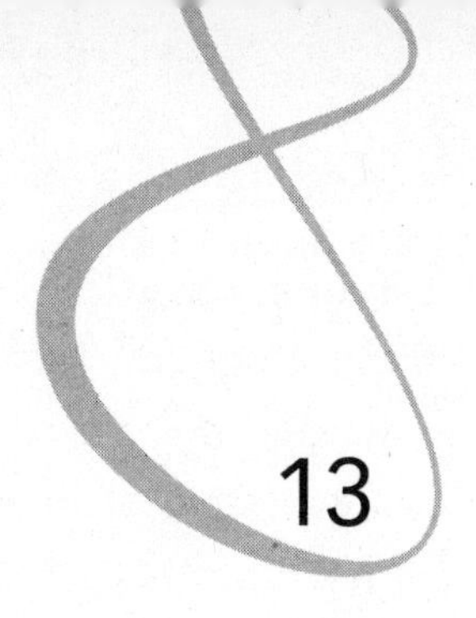

13

GOD, GÖDEL, AND THE SEARCH FOR PROOF

> I mean the word proof not in the sense of the lawyers, who set two half proofs equal to a whole one, but in the sense of a mathematician, where half proof = 0, and it is demanded for proof that every doubt becomes impossible.
>
> —CARL FRIEDRICH GAUSS

> *Proof* is an idol before whom the pure mathematician tortures himself.
>
> —ARTHUR EDDINGTON

Mathematics is the only subject in which absolute certainty is possible. Statements and theorems can be shown to be true beyond any shred of doubt, and these truths will hold for all time. That's why mathematicians are so obsessed with proof. Once something is proven, rigorously, it can be added, with total confidence, to what's already known and thereby serve as a rock-solid foundation for future research. Only one cloud hangs, permanently and frustratingly, in the otherwise clear mathematical sky: the knowledge that there will always be things that can be said, in any system of math, that can't be shown to be true or false from within that system.

In about 1941, Austrian-born logician Kurt Gödel, close friend of Albert Einstein at the Institute for Advanced Study at Princeton, proved that God exists. Unlike Einstein, who hovered between agnosticism and pantheism and once said he believed in "Spinoza's god," Gödel was a nonchurchgoing theist who, according to his wife, "read the Bible in bed every Sunday morning." The proof he published about the existence of God, however, had nothing to do with his Lutheran roots or anything that might ring a bell with ordinary folks. It was very much a product of his intellectually lofty mathematical mind. The first line of it reads:

$$\{P(\varphi) \wedge \Box \forall x[\varphi(x) \rightarrow \psi(x)]\} \rightarrow P(\psi)$$

and what follows doesn't get any clearer. It ends with the punch line:

$$\Box \exists x G(x)$$

which, for us mere mortals, translates as "something godlike necessarily exists."

Needless to say, Gödel's proof didn't go unchallenged, and although it's cloaked in the formal notation of what's called "modal logic," making it seem impressively rigorous, it involves a lot of dubious assumptions that are purely a matter of opinion. Not so with other results for which Gödel is better known,

Kurt Gödel.

most notably his world-shaking incompleteness theorems, of which more later.

"Proof" means different things to different people. In the legal profession, it comes in a number of flavors, depending on the type of case and court involved. Proof in law really boils down to evidence, the amount or quality of which, needed to satisfy a judge or jury, varies between civil and criminal cases. In civil cases a judgment is made based on the balance of probabilities: a judge can convict if he or she reaches a conclusion of "more likely than not" or "reasonable suspicion." In Anglo-American criminal cases, defendants are assumed to be innocent unless proved guilty—"proved" meaning not just probably guilty but guilty "beyond reasonable doubt."

Scientists, like lawyers, work more with evidence than proof. In fact, modern-day scientists are quite modest in their claims and avoid talking about "proof" or "truth" in any absolute sense. What science is mostly about is making observations, coming up with theories that best fit the data, and then testing these theories against more observations and experiments. Scientific theories are only ever provisional—just the best ideas at the time to explain the way the world seems to work. A single new observation, if confirmed, that goes against the theory is enough to scupper it permanently.

Take gravity, for instance. Aristotle was convinced that heavier objects fall faster than lighter ones. After all, if you drop a stone and a feather at the same time, the stone easily wins the race to the ground. It took some clever experiments—and a gap of nearly two thousand years—to show that Aristotle was wrong. There's an attractive myth that, in 1589, Galileo fatally undermined the old ideas about gravity when he climbed to the top of the Leaning Tower of Pisa, dropped two cannonballs of different weights at the same time, and noted that they hit the ground simultaneously. It probably never happened: the only primary source for it is a mention in the biography of Galileo written by one of his pupils, Vincenzo Viviani, and published well after Viviani's death. What certainly did happen is that Galileo experimented with balls of different weight rolling down inclined planes—an elegant way of diluting the effects of gravity so that he could make accurate measurements of the rates at which objects fall. Galileo's results, together with those of German astronomer Johannes Kepler, were used by Isaac Newton to come up with a new theory of gravity. It's the one still taught in schools, the one that helps mission planners to chart courses for spacecraft across the solar system, that works just fine in almost every situation where you need to know what the effects of gravity will be. Almost. The trouble is, it doesn't give accurate results in every case. Newton's universal theory of gravitation is a very good approximation—so good that we don't normally notice that its predictions differ from reality. But it's just that: an approximation. In 1915 Einstein published his general theory of relativity, which is currently our best theory of gravity. It explains things that Newton's theory can't, like the shifting orbit of Mercury, the bending of starlight as it passes near the sun, and situations where the gravitational pull is extreme, as in the vicinity of a black hole. No one for a minute believes that Einstein's general theory is the last word on gravity—it can't be, because it doesn't explain how gravity operates in the world of the extremely small, where quantum mechanics comes into play. There has to be some theory that binds quantum theory and gravity together; we just haven't managed to find it yet.

The bottom line is that although it's possible to show that a scientific theory is wrong—or, at best, a mere approximation—it's impossible to prove that it's true in all circumstances. Future discoveries, about which we know nothing, could always be waiting to torpedo the best theoretical description we can come up with today. But in mathematics, it's a completely different story.

Proof lies at the heart of all mathematics. It's not something encountered much in school, where the emphasis is on solving problems. But in higher-level math, proof is king and the ultimate goal of all researchers in the field. Mathematical theories can be proven beyond a shadow of a doubt, and once proven they never change. For example, Pythagoras's theorem, concerning the sides of right-angled triangles, has been proven with certainty: it's impossible that anyone will ever find it to be wrong, given certain assumptions, which we'll talk about in a minute. In fact, of all areas of human inquiry, mathematics and its cousin logic are unique in enabling certainty beyond doubt.

Like scientists, mathematicians may initially look for evidence of something—perhaps a rule in geometry or a pattern among numbers—before proposing a theory to unite the evidence. But unlike in science, there's no endless cycle of constantly improving the theory based on new data. No matter how many times a mathematical theory stands up to tests in different situations or using different values, its truth will never be accepted until someone comes along with a rigorous proof that can be shown to be fault free. The very fact that such proofs are possible means that mathematicians are not particularly impressed by evidence alone.

The history of proof begins in ancient Greece. Before that, mathematics was largely a practical subject, used in reckoning, building, and so on. There were rules of arithmetic and rules of thumb applied to shapes and spaces, but nothing more basic than that. Proof began to emerge in about the seventh century BC, with the activities of one of the first known natural philosophers, Thales of Miletus. Thales, whose interests spanned almost all subjects, including philosophy, science, engineering, history, and geography, proved some simple early theorems in geometry. His compatriot Pythagoras, born about a half century later, is more famous to most people because of the theorem named after him. Whether he or his followers were the first to offer some sort of proof of "Pythagoras's theorem" is impossible to say because no written records of such a proof survive from the time. The Babylonians and others knew of the existence of the rule—that the square of the longest side of a right-angled triangle equals the sum of the squares of the other two sides—and applied it in their building schemes. But who first proved it, and in exactly what form, is unknown. By later standards, it would certainly have been an informal proof. The Pythagoreans were also involved in the discovery of irrational numbers—ones that can't be expressed as one whole number divided by another. Again, the roots of the idea are hard to trace, but the myth has grown up that a member of the Pythagorean cult Hippasus proved by some means that the square root of two couldn't be expressed as a fraction. So abhorrent was this to others in the cult that they allegedly drowned Hippasus to keep the flaw in their worldview a secret. However, the few ancient sources that tell the tale of a drowning either don't mention Hippasus by name or relate that he was drowned for another offense—the blasphemy of showing that a dodecahedron can be constructed inside a sphere.

Mathematical proof took a giant step forward, and reached something like the form in which we know it today, through the work of another Greek, Euclid, who was based in Alexandria, Egypt, around the turn of the third century. In his book *Elements*, he laid the foundations for modern proof theory through his use of a combination of some basic assumptions that are taken to be self-evidently true and step-by-step reasoning, where each step, starting from one or more of the basic assumptions, is seen to follow logically and irrefutably from the previous one.

Elements deals mainly with geometry and provides, for the first time, rigorous proofs of many of the geometric theorems already known to the Greeks. Euclid starts out with five core assumptions, which became known as Euclid's postulates—for example: "A straight line segment can be drawn through any two points" and "A straight line segment can be extended indefinitely." These postulates, which today we would call axioms, are taken to be so obviously true that they don't need to be proved in themselves. And even if some proofs were offered of them, these would involve making other assumptions. The fact is, we have to start somewhere. Having set out his postulates, Euclid then reasoned line by line, each line following with watertight logic from the previous one, until he had a complete proof of some theorem or other. He could then use these theorems to prove other theorems, and so on, in a completely orderly and stepwise fashion that his readers could easily follow and check.

The geometry as set forth in *Elements*—Euclidean geometry—went largely unchallenged for more than a thousand years. But then some mathematicians started to question one of the postulates on which the great work rests. The first four of Euclid's postulates are simple, straightforward, and pretty uncontroversial, but the fifth—the so-called parallel postulate—is more complicated and not so obvious. Euclid originally stated it like this: "If a straight line falls on two straight lines in such a manner that the interior angles on the same side are together less than two right angles, then the straight lines, if produced indefinitely, meet on that side on which are the angles less than the two right angles." Later mathematicians found less convoluted ways to say the same thing; for instance, Scottish scientist John Playfair came up with this equivalent statement of the parallel postulate: "In a plane, given a line and a point not on it, at most one line parallel to the given line can be drawn through the point." The parallel postulate is also equivalent to a number of other statements, the easiest of which to understand is probably that the angles of a triangle add up to 180 degrees. But, however it's phrased, the fifth postulate seems less obvious and more contrived than the other four, and it became a popular suspicion among later mathematicians that some proof of the fifth postulate might be possible using the first four. More than a thousand years after Euclid, some Arab mathematicians began to question the very validity of the parallel postulate and provided the first hints that something might lie beyond the geometry of *Elements*.

In the first half of the nineteenth century, three mathematicians, Hungarian János Bolyai, Russian Nikolai Lobachevsky, and German Carl Gauss realized that if the parallel postulate were taken out, the result would not be a failure of Euclid's geometry but an entirely new kind of geometry. This became known as hyperbolic geometry, from the Greek for "too much" in the sense of having too much space for Euclid's flat plane. Hyperbolic geometry has constant negative curvature, which means they curve in the opposite sense to that of a sphere and at a fixed rate. In hyperbolic geometry, the angles of a triangle add up to less than 180 degrees, and Pythagoras's theorem no longer holds. This doesn't mean that Euclidean geometry is wrong and that there's a mistake with the proof of Pythagoras's theorem given by Euclid. Under the axioms that Euclid laid out, Pythagoras's theorem is proven true for all time. It's just that if these axioms are changed, then different forms of geometry arise in which different theorems apply. Replacing the fifth postulate with its negation generates a whole new geometry—hyperbolic geometry. And this same effect applies to any system of mathematics: changing the underlying axioms opens up a new mathematical realm where different rules come into play. Pythagoras's theorem can be proved to be true using the set of axioms—the five postulates—defined by Euclid. But throw out the fifth postulate, and the result is a non-Euclidean geometry in which Pythagoras's theorem is false. Mathematicians found another type of geometry that also rejects the parallel postulate but, in addition, requires that the second postulate be modified so that straight lines can't be extended indefinitely, as on the surface of a sphere. This second type of non-Euclidean geometry became known as elliptic geometry and was pioneered by German mathematician Bernhard Riemann.

Euclid showed the world how to do mathematical proof properly and precisely. He also showed how it was possible to use the same suite of axioms defined in one field to embrace all of mathematics. After *Elements* he wrote other books in which he applied his five postulates to proving various theorems outside of geometry. For example, having first recast these postulates so that they applied to number theory, he was able to prove that there are infinitely many prime numbers (numbers divisible only by themselves and 1). Modern mathematicians adopt this same approach of choosing axioms in one area of their subject that can be applied across the board, but instead of using geometry they start out from the more abstract branch of math known as set theory.

The pioneers of set theory were also—not coincidentally—the pioneers of the mathematics of infinity: Germans Georg Cantor and Richard Dedekind, whom we met in Chapter 10. Set theory came into being because of its ability to handle both finite and infinite numbers. And it does just what the label says: provide a theory of sets—collections of objects, which might be numbers, letters of the alphabet, planets, inhabitants of Paris, sets of sets, or anything else that can be dreamed of. In the world of mathematics, there's complete freedom to choose the axioms that underpin the many different forms of set theory that are possible. It just so happens that the one that most mathematicians use today, because it generally works so well, is the one called Zermelo-Fraenkel set theory. Tacked onto this is an additional special axiom known as the "axiom of choice" (AC), and the whole package is often referred to as "ZFC theory." Many of the axioms in ZFC are obvious and self-explanatory: "Two sets with the same elements are identical" and so on. But the axiom of choice is a thornier issue. In fact, it's been hailed as the most controversial axiom since Euclid's parallel postulate.

Put simply, AC says that given any collection of sets, it's always possible to choose exactly one unique member from each set to make up a new set. That seems obvious in everyday situations. For instance, one person could be picked from every country in the world and all of them put together in the same room. The trouble is, it's not obvious how to do this if there are infinitely many sets that are infinite in size. In such a situation, there may be no definite way to carry out the selection, and the axiom of choice starts to look more like an arbitrary imposition than a statement upon which everyone can agree. Having said this, most mathematicians at present are happy to accept it because it's needed for the proof of many important theorems. It also leads to some results that, at first sight, seem utterly outrageous. One of these is the Banach-Tarski paradox, or decomposition, which we met in Chapter 9 and insists that it's possible to cut a ball into finitely many parts and then rearrange the parts to make two copies of the same ball, thereby doubling the original volume. The cutting can be done only in an abstract sense—mathematically—not in real life. But still it sounds more like magic than math. Yet with the axiom of choice in place, it's possible for the intermediate pieces of the cut-apart ball to be considered as if they were disconnected clouds with no defined volume and for them to be reassembled with twice (or, for that matter, a million times) the volume they started with.

Given that mathematicians are free to choose whatever set of axioms their hearts desire and that works best for them, it seems that eventually they could choose a set that allows *any* valid statement in math to be proven from those axioms. In other words, with the right axioms in place, it ought to be possible to prove anything that is mathematically true. Leading theoreticians at the turn of the twentieth century had no reason to doubt this and actively went in search of a provably complete system of mathematics. Prominent among them was German mathematician David Hilbert, famous for many developments in modern math and for a list of what he considered to be the twenty-three most important unsolved problems at the time. In 1920 he proposed a project to show that all of mathematics stems from a correctly chosen system of axioms and that such a system can be proved to have no inconsistencies. A decade later, that ambition lay in tatters thanks to the work of Austrian (and, later, American) mathematician, logician, and philosopher Kurt Gödel.

In 1931, several years before leaving Austria and joining the Institute for Advanced Studies at Princeton, where he became a close friend of Albert Einstein, Gödel published two extraordinary and shocking theorems—his first and second incompleteness theorems. In a nutshell, the first of these theorems showed that any system of mathematics that is complex enough to include ordinary arithmetic—the kind we learn in school—can never be both complete and consistent. If a system is complete, it means that everything within it can be either proved or disproved. If it's consistent, it means that no statement can be both proved and disproved. Like a bolt out of the blue, Gödel's incompleteness theorems revealed that, in any mathematical system (apart from the very simplest), there will always be things that are true but that can't be proved to be true. The incompleteness theorems are analogous in some ways to the uncertainty principle in physics in that they expose fundamental limits to what can be known. And like the uncertainty principle, they're frustrating and inhibiting because they show that reality—including purely intellectual reality—behaves in ways that prevent us from being omniscient about everything that we try to penetrate with our minds. To put it bluntly, truth is a more powerful concept than proof, which, for a mathematician especially, is anathema.

Gödel's work, and his startling conclusions, became possible only after mathematicians and logicians recognized the need to formalize mathematical systems by underpinning them with well-defined sets of axioms. Euclid had pointed the way to this approach in the days of ancient Greece. But it was only with the development of set theory and mathematical logic, in the second half of the nineteenth century, that the formalization process could become rigorous and extended to *any* system of math imaginable. In the case of the arithmetic that we first learn about in school—arithmetic that deals with the natural numbers, 0, 1, 2, 3, . . . —an axiomatic foundation was provided by Italian mathematician Giuseppe Peano (pronounced "piano") that is still used, pretty much unchanged, by mathematicians today. Some statements in ordinary arithmetic, such as "2 + 2 = 4," seem so blindly obvious that it's hard to see why they need to be proved at all—but they do. Just because they're familiar to us from a young age, we can't just assume they can be taken for granted. In Peano arithmetic, it's straightforward to prove statements like "2 + 2 = 4," once 2 and 4 have been cast in a more generalized form, as SS0 and SSSS0, where *S* stands for the "successor" of a number. It becomes easy, too, to disprove statements such as "2 + 2 = 5" but, as you would expect, impossible to disprove "2 + 2 = 4" or prove "2 + 2 = 5." Peano arithmetic wouldn't be much use if it could handle only really basic stuff like this. Its power comes from being able to deal with much more complex statements about arithmetic, and it was originally thought by mathematicians that each and every one of these statements could be proved or disproved, given enough time. What Gödel showed, in his first theorem, was that this was in fact not the case.

As an example, he chose a particular statement about Peano arithmetic that could be neither proved nor disproved from within that arithmetic. If it can be proved, he showed, then it's false (and can be disproved), and if it can be disproved, it can also be proved—either way, Peano arithmetic, if complete, is shown to be inconsistent. We might try to take up a fallback position, by relaxing the need for completeness and asking merely for proof that Peano arithmetic, or any other system, is consistent. But the second of Gödel's incompleteness theorems knocks even this idea on its head by showing that any proof that a system is consistent (from within that system) automatically also shows the exact opposite—that it's inconsistent. Not all mathematicians, however, are convinced that, on the issue of consistency, Gödel has had the last word.

Finding a proof that the axioms of arithmetic are consistent was included by David Hilbert in 1900 as the second of his famous list of (then) unsolved problems. In 1931 Gödel seemed to scupper any hope that this was possible. But then, just a few years later, in 1936, German mathematician and logician Gerhard Gentzen, who was an assistant of Hilbert's at Göttingen between 1935 and 1939, published a paper in which he proved Peano arithmetic was consistent—on the face of it, a conclusion exactly opposite that of Gödel. Unlike Gödel, however, Gentzen didn't try to prove the consistency of Peano arithmetic from within Peano arithmetic. Instead, he resorted to the properties of certain ordinals and, in particular, one very large ordinal—which we ran into earlier, in Chapter 10—that Cantor had named epsilon-zero (ε_0). So colossal is this number that Peano arithmetic can't describe it. Yet as Gentzen discovered, it can be used to express and prove statements that Peano arithmetic can't prove—in particular, Peano arithmetic's own consistency.

Gentzen's methods can be extended to prove the consistency of many systems, provided that a sufficiently large ordinal can be constructed. In fact, it turns out, every mathematical system has a certain "ordinal strength" that determines what ordinals the system can and can't express. For example, the ordinal strength of Peano arithmetic is ε_0, which means that Peano arithmetic can express any ordinal below epsilon-zero but not epsilon-zero itself. Larger, more encompassing systems have larger ordinal strengths. In the case of ZFC, the ordinal strength is unknown. What *is* known, thanks to Gentzen, is that ZFC can be augmented with certain axioms, called "large cardinal axioms," to describe cardinals that are far beyond anything ZFC can express, resulting in even stronger systems with larger (but again unknown) ordinal strengths.

Mathematicians are still divided over the issue of Hilbert's second problem: to find a proof that arithmetic is consistent. Some favor Gödel's negative solution—that it's impossible ever to obtain such a proof—while others lean toward Gentzen's partial positive proof. In any event, the question doesn't impact the central message of Gödel's theorems, which is that, working from *within* a given system of math (such as Peano arithmetic or ZFC), certain statements can be made that are undecidable. We may be able to reason about a system from a different system in order to prove or disprove these statements (as Gentzen did, considering a simple form of arithmetic that was augmented by ordinals), but we still wouldn't know if *that* system was consistent unless we simply accepted it.

For three decades after the incompleteness theorems were published in the early 1930s, few examples of undecidable statements were known, apart from highly contrived ones such as that used in Gödel's own proof. Then came a major breakthrough, concerning an idea that had been bothering mathematicians ever since it was hatched, by Cantor, in 1873. This idea was the continuum hypothesis, which we encountered in Chapter 10. CH claims that aleph-one ($\aleph_1$)—the cardinality of the set of countable ordinals—is the same as the cardinality of the set of real numbers, which is to say that there are as many real numbers (or points on a line) as countable ordinals. If CH were true, then there would be no set with a cardinality between that of the whole numbers and that of the reals. Cantor himself couldn't prove this, despite struggling for much of his life in the attempt and perhaps thereby contributing to his later mental instability. So important did Hilbert rate it that he listed it as the very first of his twenty-three great problems. Not until 1963, through the work of American mathematician Paul Cohen, was the status of CH clarified—if not completely settled. Cohen showed that from within the confines of ZFC (and they are not that confining!), the most widely used axiomatic basis for modern mathematics, the continuum hypothesis is undecidable. He found that it's possible to come up with two different sets of axioms, both containing all the axioms of ZFC and consistent in themselves, in one of which the continuum hypothesis is true and in the other it's false. Simply put, from inside ZFC, we can both prove and disprove CH, depending on which additional rules we choose. In ZFC without additional axioms, neither is possible.

Even within the much simpler mathematics of Euclid, this kind of undecidability crops up, as we've seen. Many of Euclid's earlier theorems, including all of his first twenty-eight propositions, make no use of his fifth postulate—the one about parallel lines never meeting. These theorems belong to a system that's become known as "absolute geometry"—geometry based on the axiom system for Euclidean geometry with the fifth postulate taken out. In absolute geometry, Pythagoras's theorem is undecidable because in Euclidean geometry it is true, whereas in non-Euclidean geometry, which is also based on Euclid's axioms but without the parallel postulate, such as hyperbolic geometry, it's false. Likewise, there are axioms such as those known as forcing axioms that, if added to ZFC, allow the continuum hypothesis to be disproved and other axioms such as the inner-model axiom that, if added to ZFC, allow the continuum hypothesis to be proved. The bottom line is that, given current methods, CH is provably unsolvable. Even with the tools of modern set theory, which are so powerful that they cover all of existing mathematics, CH can't be solved. Yet mathematics continues to evolve and expand, and there remains hope that through the use of new techniques, such as large cardinal axioms, a solution might be forthcoming.

The most famous claim in mathematics that (until recently) lacked a proof is Fermat's last theorem. It's not a very good name because it was neither the last theorem that French mathematician Pierre de Fermat worked on nor, at the time he proposed it, a theorem at all. Older works more accurately refer to it as Fermat's conjecture. The reason it's called his "last" theorem is that it was discovered thirty years after his death, by Fermat's son Samuel, scribbled in the margin of a book from Pierre's collection—*Arithmetica*, by Diophantus. Fermat's claim is quite easy to state: There are no integer solutions to the equation $x^n + y^n = z^n$ for values of n greater than 2. If n equals 2, then there are infinitely many solutions—for example, $3^2 + 4^2 = 9 + 16 = 25 = 5^2$. But there are no solutions at all, Fermat insisted, if n is 3 or more. "I have a truly marvelous demonstration of this proposition," he wrote (in Latin), "which this margin is too narrow to contain."

Now, Fermat was a great mathematician, not prone to error. No mistakes have been found in any of the proofs he published. The only conjecture he ever made that was later disproven is one for which he never claimed to have a proof. Was he joking

Engraving of Pierre de Fermat.

in his cryptic comment? Was this his way of challenging contemporary and future mathematicians to come up with a proof? Or was he stating a fact when he said he had a proof but not the space to write it down? History suggests it wasn't the last of these because, despite numerous efforts, no one in the centuries that followed was able to come up with a reasonably short proof. In fact, it wasn't until 1995, 358 years after Fermat penned his tantalizing note, using mathematics that was vastly more advanced than anything available in the seventeenth century, that Fermat's conjecture was finally elevated to the status of a proven theorem.

Credtit for solving the problem went to British mathematician Andrew Wiles, who had been fascinated by Fermat's claim ever since first reading about it in a book in his local library on the way home from school at the age of ten. Almost a quarter of a century later, he began in earnest to look for a proof—a quest that took him into an area of math connected with elliptic curves and a proposition known as the Taniyama-Shimura conjecture that had been formulated by Japanese mathematicians Yutaka Taniyama and Goro Shimura in 1957. Wiles announced a proof of Fermat's last theorem during a lecture in 1993, but subsequently this proof was found to have a mistake, and it was only two years later, after he had almost given up trying to fix the mistake, that Wiles finally put forward a flawless proof that settled the matter once and for all. While Fermat's last theorem is one of the most famously difficult mathematical problems, it really isn't that important to mathematicians. It wasn't included, for instance, in Hilbert's problems. On the other hand, the Taniyama-Shimura conjecture has important results, linking what appear to be vastly different fields of mathematics.

Proofs like that of Fermat's last theorem are hard because they're complex and require truly inspirational breakthroughs. Others are hard mainly because they're laborious and enormously time-consuming. The so-called four-color theorem, which states that any map can be colored using just four colors so that no two adjacent regions have the same color, first cropped up in a letter from Augustus De Morgan, first professor of mathematics at the new University College London, to his friend Irish mathematician William Hamilton in 1852. Restrictions on the problem are that each region of the map must be connected, the regions must lie on the plane, and any two regions must actually share part of a border to be connected—a single point doesn't count. It turns out that this is really hard to prove. In terms of the theory alone, it's no pushover, but the biggest problem is the number of different possibilities that have to be checked. Eventually, after more than a century of effort, mathematicians, having considered all the ways maps could be drawn, had reduced the number of unique configurations to 1,936. Even this was far too many for a human or a team of humans to check in a lifetime, and so computers were used to do the number crunching. In 1976 the four-color theorem was finally proved by Kenneth Appel and Wolfgang Haken at the University of Illinois and double-checked using different programs and computers.

Despite the care used by Appel and Haken to cross-check their result, there was an outcry from some mathematicians and philosophers who argued that proof by machine was either not legitimate or not reliable, as it was impossible for humans to verify by hand. This debate about the use of computers to prove theorems rumbles on, amid concerns that results may be wrong if a computer malfunctions or there's an error in the software it runs. But out of necessity, it's an approach that's becoming more common and acceptable as time goes on. A recent development that offers some reassurance to skeptics is the rise of "computer proof assistants," which are programs that can format proofs and proofread them for errors.

An area of math rife with the need for monstrously long proofs is Ramsey theory, the gist of which is that if you color the members of any set, it's impossible to avoid certain patterns from appearing. One of the problems in Ramsey theory goes by the name of "Boolean Pythagorean triples." The problem asks if it's possible to color each of the positive integers either red or blue, so that no Pythagorean triple of integers a, b, c, satisfying $a^2 + b^2 = c^2$, are all the same color. In May 2016, Marijn Heule, Oliver Kullmann, and Victor Marek ran one of the fastest computers in the world, Stampede, at the Texas Advanced Computing Center in Austin, for two days and showed, in a two-hundred-terabyte proof, that such a coloring is impossible. To get some idea for how long the proof is, it's been estimated that it would take a person around 10 billion years (roughly the lifetime of the sun) just to read it—and a great deal longer to verify it. Even longer proofs are likely in the future. One possible candidate is Ramsey's theorem for $n = 5$. It's known that given forty-nine vertices of a graph, if the edges are colored in one of two different colors, there's guaranteed to be five vertices such that all edges between them are of the same color. It's also known that this is not true for forty-two vertices, but finding a proof of the minimum number is a challenge for mathematicians armed with still greater computational power.

* * *

Mathematics, contrary to how it may sometimes seem, is an endless adventure into the weirdest and wildest places ever countenanced by the human intellect. It lures us into thinking that it's ordinary and banal, because its roots are in the familiar—in simple numbers and shapes. It began as the tool of the merchant, the farmer, the builder of temples and pyramids, the early watchers of the seasons and the skies. But it's anything but ordinary. It permeates every aspect of the reality in which we're embedded, forming an invisible infrastructure behind the behavior of everything around us, from the smallest particle to the universe as a whole.

For the most part, we live our lives in the belief that what we see and experience every day is normal and unremarkable. Yet it's anything but that. We're built of atoms whose nuclei, for the most part, were fused in the cores of giant stars; we're almost literally formed of stardust. So when we look up at the night sky, we're seeing the place from which, ultimately, we came. Our day-to-day existence depends on sunlight captured by chemicals in organisms that evolved from simpler creatures that somehow sprang into being on the surface of a young and desolate world. All of the spacetime around us emerged spontaneously from an unimaginably tiny point some 14 billion years ago and is now hurtling toward a future that we have yet to determine. Ninety-five percent of the matter and energy in the universe are in the forms of dark matter and dark energy, whose nature remains mysterious. And all of this extraordinary activity and unfolding, from the submicroscopic to the cosmic in scale, is guided by the unseen hand of mathematics.

On occasions we find that some aspect of math, developed for its own sake, without any thought as to whether it might eventually prove useful, turns out to describe with fantastic precision how materials behave under certain conditions or what happens to subatomic particles when they collide at close to the speed of light. Outrageous excursions into convoluted topologies, higher dimensions, and fractal landscapes find practical application in technology, physics, chemistry, astronomy, and music. The very beating of our heart, the convolutions of our lungs, the firing of our synapses with every thought—including those we're having this very instant—are guided by equations and patterned by mathematical logic.

We may sometimes think that math is detached from the real world, but it's here, right now, present in everything we see and do. We may sometimes feel, too, that our lives are routine and mundane. Yet the fact is that we're at the center of something remarkable, and behind all of this extraordinary riot of creation are the wonder and the weirdness of math.

ACKNOWLEDGMENTS

Our thanks go to Agustin Rayo of MIT, Adam Elga of Princeton University, Winfried Hensinger of the University of Sussex, and Andrew Barker for reading through parts of the manuscript and offering advice. We are also grateful to our editors, T. J. Kelleher, Carrie Napolitano, and Hélène Barthélemy at Basic Books, and, in the United Kingdom, Sam Carter and Jonathan Bentley-Smith at Oneworld, not to mention all of the other talented folks in design and production who have helped bring this book to fruition.

Agnijo would like to thank Ms. Hannah Young, Mrs. Yvonne O'Brien, and Ms. Helen Treece—three teachers who have always inspired and motivated him—and all the staff of Grove Academy, Broughty Ferry, Scotland, for their encouragement. Most of all, his thanks go to his mum, dad, and younger brother for their support in everything.

As always, David has relied on the love and support of his ever-patient wife, Jill, and his children and grandchildren. He will always be grateful, too, to his parents for everything they did.

INDEX

ABOUT THE AUTHORS

DAVID DARLING is a science writer with a PhD in astronomy. He is the author of many books, including the best-selling *Equations of Eternity* and *Mayday! A History of Flight Through Its Martyrs, Oddballs and Daredevils.* He lives in Dundee, Scotland.

AGNIJO BANERJEE is a brilliant young mathematician and child genius, with an IQ of at least 162, according to Mensa. He has just finished training in Hungary in preparation for the 2017 International Mathematics Olympiad. A student of Darling's, he lives near Dundee, Scotland.